# Science Activity Workbook

**Grade 5**

Macmillan/McGraw-Hill

The McGraw·Hill Companies

 **Macmillan
McGraw-Hill**

Published by Macmillan/McGraw-Hill, of McGraw-Hill Education, a division of The McGraw-Hill Companies, Inc., Two Penn Plaza, New York, New York 10121. Copyright © by Macmillan/McGraw-Hill. All rights reserved. The contents, or parts thereof, may be reproduced in print form for non-profit educational use with Macmillan/McGraw-Hill Science, provided such reproductions bear copyright notice, but may not be reproduced in any form for any other purpose without the prior written consent of The McGraw-Hill Companies, Inc., including, but not limited to, network storage or transmission, or broadcast for distance learning.

Printed in the United States of America

4 5 6 7 8 9 024 09 08 07 06 05

# Table of Contents

**Unit A — Characteristics of Living Things**
Chapter 1—Classifying Living Things . . . . . . . . . . . . . . . . . . . . . . . . . . . . . . . . . . . . . 1
Chapter 2—Plant Structure and Functions . . . . . . . . . . . . . . . . . . . . . . . . . . . . . . . . 11
Chapter 3—Plant Diversity . . . . . . . . . . . . . . . . . . . . . . . . . . . . . . . . . . . . . . . . . . . . 21
Chapter 4—Animal Diversity . . . . . . . . . . . . . . . . . . . . . . . . . . . . . . . . . . . . . . . . . . 36

**Unit B — Living Things and Their Environment**
Chapter 5—Interactions of Living Things . . . . . . . . . . . . . . . . . . . . . . . . . . . . . . . . 46
Chapter 6—Ecosystems . . . . . . . . . . . . . . . . . . . . . . . . . . . . . . . . . . . . . . . . . . . . . . 62

**Unit C — Earth and Its Resources**
Chapter 7—Landforms, Rocks, and Minerals . . . . . . . . . . . . . . . . . . . . . . . . . . . . . 79
Chapter 8—Air, Water, and Energy . . . . . . . . . . . . . . . . . . . . . . . . . . . . . . . . . . . . 100

**Unit D — Astronomy, Weather, and Climate**
Chapter 9—Astronomy . . . . . . . . . . . . . . . . . . . . . . . . . . . . . . . . . . . . . . . . . . . . . 120
Chapter 10—Weather . . . . . . . . . . . . . . . . . . . . . . . . . . . . . . . . . . . . . . . . . . . . . . 132
Chapter 11—Weather Patterns and Climate . . . . . . . . . . . . . . . . . . . . . . . . . . . . . 153

**Unit E — Properties of Matter and Energy**
Chapter 12—Properties and Structure of Matter . . . . . . . . . . . . . . . . . . . . . . . . . . 170
Chapter 13—Forms of Matter and Energy . . . . . . . . . . . . . . . . . . . . . . . . . . . . . . . 185

**Unit F — Motion and Energy**
Chapter 14—Newton's Laws of Motion . . . . . . . . . . . . . . . . . . . . . . . . . . . . . . . . . 208
Chapter 15—Sound Energy . . . . . . . . . . . . . . . . . . . . . . . . . . . . . . . . . . . . . . . . . . 223
Chapter 16—Light Energy . . . . . . . . . . . . . . . . . . . . . . . . . . . . . . . . . . . . . . . . . . . 238

Name_____ Date_____

**Explore Activity**
**Lesson 1**

# What Is the Basic Unit of Life?

**Hypothesize** Most plants live on land, but some live in water. Some are tiny, and others grow very large. Do all plants have common traits?

Write a **Hypothesis:**

_____
_____

## Materials

- *Elodea* plant, moss, fern, or any flowering plant
- prepared slide of human blood
- microscope
- microscope slide
- coverslip
- dropper
- water

## Procedure

1. **Observe** Your group will need to get a plant from your teacher and a prepared slide of human blood.

2. **Communicate** As you observe each plant, draw the plant and describe it.

_____
_____
_____
_____

3. Make a wet-mount slide of your plant by placing a leaf in a drop of water in the center of the slide and carefully putting a coverslip on top.

4. **Observe** View the slide under low power. Then observe the prepared slide of human blood.

_____

5. **Communicate** Draw what you see.

Unit A · Characteristics of Living Things        Use with textbook page A5

Name _____ Date _____

**Explore Activity**
**Lesson 1**

## Drawing Conclusions

1. **Communicate** What common traits did you observe using the microscope?

   _____
   _____
   _____

2. **Communicate** What do the organisms the cells come from have in common?

   _____
   _____
   _____

3. **Define** From what you observed, come up with your own definition of a living thing.

   _____
   _____
   _____

4. **FURTHER INQUIRY** **Hypothesize** Examine the drawings you made and think about the organisms that they come from. Do you see any differences? Why do you think cells vary form one organism to another.

   _____
   _____
   _____
   _____
   _____

**Inquiry**

Think of your own questions that you might like to test. What traits are common in plant cells?

My Question Is:

_____

How I Can Test It:

_____

My Results Are:

_____

Name_____  Date_____

# A Look at Plant Cells

**Procedure**

1. Look at prepared slides of a variety of plant cells. Describe what you see. Then make a drawing of each kind of cell.

**Materials**
- prepared slides of plant cells
- various colors of clay
- toothpicks
- pieces of colored paper

2. Use clay and other materials to make a model of a plant cell. Be sure to show inside structures.

**Drawing Conclusions**

1. How were the cells you looked at similar?
   _____
   _____
   _____

2. How were the cells you looked at different from each other?
   _____
   _____
   _____

3. How did you show different plant parts in your model?
   _____
   _____

Unit A · Characteristics of Living Things          Use with TE textbook page A5

Name_____ Date_____

# Plant Parts

**Hypothesize** How does water get to different parts of a plant? Write a **Hypothesis**:

_____
_____

## Procedure

1. **Observe** Use a hand lens to observe the parts of a celery plant.

**Materials**
- celery stalk
- water
- food coloring
- narrow-mouthed bottle
- hand lens
- knife

2. Draw what you see.

3. **Hypothesize** Make a guess about the function of the stalk of the celery plant.
_____

4. Label the plant organs you see on your drawing. What levels of organization does your drawing show?
_____
_____

Unit A · Characteristics of Living Things        Use with textbook page A9

Name_____ Date_____

**QUICK LAB**
FOR SCHOOL OR HOME
Lesson 1

5. **Experiment** Add water to a bottle so that the water is about 1 inch deep. Add a few drops of food coloring to the water. Cut a piece of stalk and place it in the colored water. Observe and then write what you see after a few minutes.

## Drawing Conclusions

6. **Communicate** Explain to the class why your observation supports or doesn't support your guess.

7. **Going Further** The tubelike structure, or vascular tissue, in a celery plant transports water very well. How can you demonstrate water moving up a tube? Write and conduct an experiment.

My Hypothesis Is:

_____

_____

My Experiment Is:

_____

_____

_____

My Results Are:

_____

_____

Unit A · Characteristics of Living Things          Use with textbook page A9

Name _____  Date _____

# Explore Activity
## Lesson 2

# What Traits Are Used to Classify Plants?

**Materials**
- lettuce leaf
- liverwort plant
- stalk of celery
- hand lens

**Hypothesize** How do plants differ from each other?

Write a **Hypothesis:**

_____

_____

_____

## Procedure

1. **Observe** Use the hand lens to observe the lettuce leaf and the liverwort.
2. **Communicate** As you observe each plant, draw and describe the plant.

3. **Observe** Break the piece of celery. Pull apart the two pieces. Remove a 1-cm piece of the string from the celery. Observe the strings with the hand lens.
4. **Communicate** Draw and describe the string from the celery.

Unit A · Characteristics of Living Things    Use with textbook page A13

Name_____ Date_____

# Explore Activity
## Lesson 2

## Drawing Conclusions

1. **Communicate** How are the lettuce, liverwort, and celery similar?
   _____

2. **Communicate** How are the plants different?
   _____
   _____

3. **Infer** What is the purpose of the tubelike parts in the lettuce and the celery?
   _____

4. **Infer** How can liverworts live and grow without tubelike parts?
   _____

5. **FURTHER INQUIRY** **Make a Model** Design a model to show how both kinds of plants get water and minerals from the soil.

## Inquiry

Think of your own questions that you might like to test. How do plants adapt to the change of seasons?

My Question Is:
_____

How I Can Find Out:
_____
_____

My Results Are:
_____

Unit A · Characteristics of Living Things        Use with textbook page A13        7

Name_____ Date_____

# Plant Parts

**Procedure**

1. Observe the different stems and describe how they are different.
   _____
   _____

2. Observe the different leaves and describe how they are different.
   _____
   _____

3. Observe the different roots and describe how they are different.
   _____
   _____

**Materials**
- stems of different lengths
- leaves of different sizes and shapes
- taproots and fibrous roots

**Drawing Conclusions**

1. For each stem that you looked at, predict where the plant might live and how its stem helps the plant survive.
   _____
   _____

2. For each leaf that you looked at, predict where the plant might live and how its leaf helps the plant survive.
   _____
   _____

3. For each root that you looked at, predict where the plant might live and how its root helps the plant survive.
   _____
   _____

Name_____ Date_____

**Inquiry Skill Builder**
**Lesson 2**

# Classify

## Using a Key

How should a living thing be classified? Into what group should it be placed?

One way to classify organisms is by using a *classification key*. A classification key lists choices describing characteristics of organisms. It is a series of pairs of statements with directions to follow. These directions will eventually lead you to the identity of the organism you have chosen.

## Procedure

1. **Observe** Use the classification key to identify the birds shown on page A20 of your textbook. Starting with the first pair of statements, choose the one that applies to the bird you picked.

2. **Interpret Data** Follow the statement's directions. It will lead you to another pair of statements.

3. Keep following the directions until you come to the identity of the bird you chose.

### Key to Birds

1. Webbed feet . . . . . . . . . . . . Go to 3.
   No webbed feet . . . . . . . . . Go to 2.

2. Hooked bill . . . . . . . . . . . . . Red-tailed Hawk
   No hooked bill . . . . . . . . . . Cardinal

3. Flat bill . . . . . . . . . . . . . . . . Mallard Duck
   No flat bill . . . . . . . . . . . . . Go to 4.

4. Pouch . . . . . . . . . . . . . . . . . Brown Pelican
   No pouch . . . . . . . . . . . . . . Red-faced Cormorant

Unit A · Characteristics of Living Things    Use with textbook page A20

Name _____ Date _____

**Inquiry Skill Builder**
**Lesson 2**

## Drawing Conclusions

Do you think this key would be helpful in identifying birds in your neighborhood? Explain.

_____

_____

Unit A · Characteristics of Living Things

Use with textbook page A20

Name _____ Date _____

## Explore Activity
### Lesson 3

# How Do a Plant's Parts Help It Survive?

**Materials**
- cactus
- water plant, such as an *Elodea* or a duckweed
- flowering plant, such as a geranium

**Hypothesize** How may plants from different places differ from each other? How do the differences help the plants survive in their surroundings?

Write a **Hypothesis:**

_____
_____

## Procedure

1. **Observe** Look at the physical properties of the leaves of each plant. Note the color, size, and shape of the leaves.

   _____
   _____

2. **Communicate** List any other plant parts that you see.

   _____
   _____

3. **Communicate** Observe the physical properties of these parts and record your observations.

   _____
   _____

Unit A · Characteristics of Living Things          Use with textbook page A29

Name_____ Date_____

**Explore Activity**
Lesson 3

## Drawing Conclusions

1. **Interpret Data** How do the parts of a cactus help it survive in a hot, dry desert?

   _____
   _____

2. **Infer** Would the geranium be able to survive in the desert? Why or why not?

   _____
   _____

3. **Infer** Could the water plant survive out of water? Why or why not?

   _____

4. **FURTHER INQUIRY** **Predict** Could these plants survive outdoors where you live? Why or why not? For each plant, what conditions would you have to change so that the plant could survive outside where you live?

   _____
   _____
   _____
   _____

**Inquiry**

Think of your own questions that you might like to test. How do plants adapt to the change of seasons?

My Question Is:

_____

How I Can Find Out:

_____
_____

My Results Are:

_____

Unit A · Characteristics of Living Things            Use with textbook page A29

Name_____ Date_____

**Alternative Explore**
**Lesson 3**

# In the Dark

**Procedure**

1. Obtain two similar plants.
2. Place one plant in a dark area, such as a closet. Place the other plant in a sunny area.
3. Observe the plants each day over the course of two weeks and record your observations. Remember to give both plants the same amount of water.

**Materials**
- two similar plants

| Day | Plant in a Sunny Place | Plant in a Dark Place |
|---|---|---|
|  |  |  |
|  |  |  |
|  |  |  |
|  |  |  |
|  |  |  |
|  |  |  |
|  |  |  |
|  |  |  |
|  |  |  |
|  |  |  |
|  |  |  |

**Drawing Conclusions**

1. After two weeks, how did the plants look?

   _____

   _____

2. What do you think will happen to the plant that was in the dark if you put it in the light?

   _____

   _____

Unit A · Characteristics of Living Things        Use with TE textbook page A29

Name_____ Date_____

# Leaves

**QUICK LAB FOR SCHOOL OR HOME**
**Lesson 3**

**Hypothesize** In what ways are the leaves that are important to you alike? In what ways are they different? Write a **Hypothesis:**

_____

_____

_____

**Materials**

- various plant leaves that you eat
- hand lens

## Procedure

1. Collect a variety of different leaves that you eat as food.
2. **Observe** Examine them with a hand lens.

## Drawing Conclusions

3. Make a sketch of at least 12 leaves, including their veins, on quarter sheets of photocopy paper.
4. **Classify** Into how many kinds of vein patterns can you group your sketches? Use quarter sheets of notebook paper to explain similarities and differences you used to classify the leaves.

_____

_____

Unit A · Characteristics of Living Things                Use with textbook page A35

Name_____ Date_____

**Quick Lab FOR SCHOOL OR HOME — Lesson 3**

5. **Going Further** What parts of plants are vegetables? What parts of plants do people eat? Write and conduct an experiment.

My Hypothesis Is:

_____

_____

My Experiment Is:

_____

_____

My Results Are:

_____

_____

_____

_____

_____

_____

| Root | Stem | Bud | Fruit | Seed |
|------|------|-----|-------|------|
|      |      |     |       |      |
|      |      |     |       |      |
|      |      |     |       |      |
|      |      |     |       |      |
|      |      |     |       |      |
|      |      |     |       |      |

Unit A · Characteristics of Living Things · Use with textbook page A35

# How Do Roots Grow?

**Explore Activity — Lesson 4**

**Hypothesize** Do roots always grow "down" no matter how you plant a seed?

Write a **Hypothesis**:

_____

## Materials
- petri dish (plastic)
- 2 paper towels
- marking pen
- tape
- 4 bean seeds that have been soaked in water overnight

## Procedure

1. Soak two paper towels. Wrinkle the paper towels and place them in the bottom half of the petri dish.

2. Place the four seeds on top of the wet paper towels as shown in diagram 1. Place the seeds so the curved part is turned toward the center of the dish.

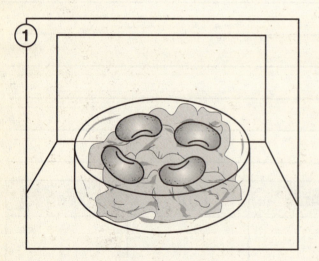
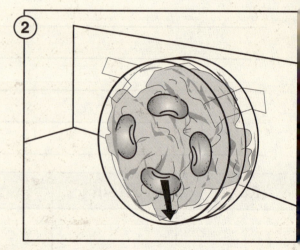

3. Place the top on the petri dish. The top will hold the seeds in the wet paper towels. Seal the top with transparent tape. Draw an arrow on the petri dish with the marking pen as shown in diagram 2. This will show which direction is down. Write the number or name of your group on the petri dish.

4. In a place your teacher provides, stand the petri dish on its edge so the arrow is pointing downward. Tape the petri dish so that it will remain standing. Do not lay the dish down flat.

5. **Predict** Make and record a prediction about the direction you think the roots will grow.

_____

Name_____ Date_____

**Explore Activity Lesson 4**

6. **Communicate** Examine the seeds for the next four days. Record the direction of root growth.

_____
_____

## Drawing Conclusions

1. **Observe** In what direction were the roots growing on day 1? On day 4?

_____
_____

2. **Interpret Data** Is your prediction supported by your data?

_____

3. **FURTHER INQUIRY** **Predict** What would happen if a seedling were not able to grow its roots down into the soil? Design an experiment to test your prediction.

_____
_____
_____

**Inquiry**

Think of your own questions that you might like to test. What happens if a germinated seed is disturbed so that the orientation of the roots is changed?

My Question Is:

_____
_____

How I Can Test It:

_____
_____

My Results Are:

_____

Name _____ Date _____

Lesson 4

# Which Way Do Corn Plant Roots Grow?

**Procedure**

1. Line a clear plastic cup with damp paper towels.
2. Place 5 soaked corn seeds between the paper towel and the side of the cup.
3. The seeds should be halfway between the bottom and the rim of the cup. Turn the seeds so that the pointed end of each one points in a different direction.
4. Use the marking pen to number the seeds.
5. Observe the growth of roots over the next few days. In the table, make drawings of each seed each day. Note the direction of the roots.

**Materials**

- presoaked corn seeds
- marking pen
- clear, tall plastic cups
- paper towels

| Day | Seed 1 | Seed 2 | Seed 3 | Seed 4 | Seed 5 |
|---|---|---|---|---|---|
|  |  |  |  |  |  |
|  |  |  |  |  |  |
|  |  |  |  |  |  |
|  |  |  |  |  |  |

**Drawing Conclusions**

1. At the beginning, in what direction did the roots grow?
   _____

2. After a few days in what direction did the roots grow?
   _____

3. What can you conclude about how roots grow?
   _____

Name_____ Date_____

**Inquiry Skill Builder** — Lesson 4

# Experiment

**Why Leaves Change Color**

To find out why leaves change color in autumn, the first thing you might do is figure out what changes occur in the fall that might cause leaves to change color. Scientists call such changes *variables*. You might identify two of these variables as the amount of daylight and the temperature, both of which go down in the fall.

Next you would make a guess that seems to make sense about which variable causes leaves to change color. This guess is called a *hypothesis*. It is often made in the form of an *if . . . then . . .* statement. For example, "*If* the plant doesn't get water, *then* it won't grow." To see if your hypothesis is a good idea, you would perform an experiment. That experiment has to be set up so that it gives a clear answer.

A

B

C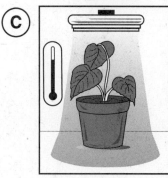

## Procedure

1. Look at the drawings. They show three experiments—A, B, C. Study the setups.

2. **Observe** What variable or variables are being tested in the first experiment? Record your answer. What variable or variables are being tested in the other two experiments?

   _____

   _____

Name_____ Date_____

**Inquiry Skill Builder**
**Lesson 4**

## Drawing Conclusions

1. **Infer** Which experiment is testing to see whether light causes leaves to change color? Explain.

   _____
   _____

2. **Infer** Which experiment is testing to see whether temperature causes leaves to change color? Explain why.

   _____
   _____

3. **Infer** Which experiment will not give a clear answer? Explain why not.

   _____
   _____
   _____

20    Unit A · Characteristics of Living Things    Use with textbook page A48

Name_____ Date_____

# What Are the Parts of Mosses?

**Explore Activity**
**Lesson 5**

**Hypothesize** Why do ferns grow tall while mosses grow only very close to the ground? How do the parts of mosses help them live where they do?

Write a **Hypothesis:**

_____
_____
_____
_____
_____

**Materials**
- hand lens
- forceps
- dropper
- 3 microscope slides
- coverslip
- microscope
- moss plant

## Procedure

1. **Observe** Place a moss on a paper towel. Use a hand lens to find its rootlike, stemlike, and leaflike parts. Record your observations.

   _____
   _____

2. **Measure** Use the forceps to remove a leaflike part. Make a wet-mount slide of the part. Observe its cells using the microscope on low power. Determine how thick the leaflike part is by moving the focus up and down.

3. **Observe** Find a capsule-shaped object at the end of the brownish stalk. Observe it with the hand lens. Place the capsule on a slide. Add a drop of water. Place a second slide on top of the capsule. Press down on the top slide with your thumb and crush the capsule. Carefully remove the top slide and place a coverslip over the crushed capsule. Examine the released structures under low power. On a separate sheet of paper, draw what you see.

Name _____ Date _____

**Explore Activity — Lesson 5**

**Drawing Conclusions**

1. **Observe** Which parts of the moss are green? Explain why they are green.

   _____
   _____

2. **Observe** How many cell layers make up the leaflike structure?

   _____
   _____

3. **Interpret Data** What structures anchor the moss plant? What was the capsule?

   _____
   _____

4. **FURTHER INQUIRY** **Predict** What do you think the objects inside the capsule do? How would you set up an experiment to test your prediction?

   _____
   _____

**Inquiry**

Think of your own questions that you might like to test. How do cells of mosses compare with other plant cells?

My Question Is:
_____
_____

How I Can Test It:
_____
_____

My Results Are:
_____
_____

22  Unit A · Characteristics of Living Things    Use with textbook page A57

Name_____ Date_____

## Alternative Explore
**Lesson 5**

# Parts of Mosses

## Procedure

1. Your teacher will give your group some pictures of mosses.

2. In the pictures, find the rootlike hairs, the stemlike part, and the leaflike part. Record your observations.

   _____
   _____

3. Look at illustrations of moss cells and the inside of a spore capsule. Draw what you see.

### Materials

- pictures and diagrams of different mosses and parts of mosses

## Drawing Conclusions

1. Is there anything in the pictures of moss plants to tell you how large the plants are? If so, describe their size.

   _____

2. Which part of a moss plant looks like it lasts longer: the green part or the spore case? Explain.

   _____
   _____

Name _____ Date _____

# Ferns

**FOR SCHOOL OR HOME**
**Lesson 5**

**Hypothesize** In what ways are ferns and mosses alike and different? Examine a fern and compare the results to those from the Explore Activity. Write a **Hypothesis:**

_____

_____

### Materials

- fern plant
- fern leaf with spore cases
- microscope
- microscope slide
- toothpick
- water

## Procedure

1. **Observe** Carefully examine the whole fern plant. Look at the stem. Observe how the leaves grow from the stem. Find veins in the leaves. Draw what you see below.

2. **Observe** Find a leaf whose bottom is covered with brownish spots. These are spore cases.

3. **Experiment** Place a drop of water on a clean slide. Use a toothpick to scrape one of the spore cases into the drop of water.

4. **Observe** Examine the spore case under the low power of a microscope. What does the spore case contain?

_____

24  Unit A · Characteristics of Living Things     Use with textbook page A60

Name_____ Date_____

**Quick Lab** — FOR SCHOOL OR HOME — Lesson 5

## Drawing Conclusions

5. **Infer** What do fern and mosses have in common?

   _____

6. **Going Further** What is the function of fern spores? How can you demonstrate this? Write and conduct an experiment.

   My Hypothesis Is:

   _____
   _____

   My Experiment Is:

   _____

   My Results Are:

   _____

| Name | Date |

# How Do Seed Plants Differ?

**Explore Activity**
Lesson 6

**Hypothesize** Have you ever noticed the differences in plant leaves? Are some leaves larger than others? How do these differences help the plant survive?

Write a **Hypothesis:**

_____

_____

## Materials

- small pine seedling or other conifer
- grass plant
- garden plant or house plant, such as geranium
- hand lens
- microscope slide
- coverslip
- microscope

## Procedure

1. **Observe** Examine each plant. Use the hand lens to examine a leaf from each one. On a separate piece of paper, draw each leaf, and label it with the name of the plant it came from.

2. **Observe** Remove a part of the lower epidermis from the grass leaf. Make a wet-mount slide. Examine the slide under low power.

3. **Communicate** On a separate piece of paper, draw what you observe.

4. **Observe** Repeat step 2 with a pine needle and a houseplant leaf (such as a geranium). On a separate piece of paper, draw what you observe.

## Drawing Conclusions

1. **Interpret Data** How are the leaves of the three plants alike? How are the leaves of the three plants different from one another?

_____

_____

_____

_____

Unit A · Characteristics of Living Things       Use with textbook page A67

Name_____ Date_____

**Explore Activity — Lesson 6**

2. **Infer** Which one of the plants do you think is least like the other two? Explain your reasoning.

_____
_____

3. **FURTHER INQUIRY** **Experiment** Predict which of the plants you examined could survive best in a dry environment. How do you think the plant's leaves would help it do this? Design an experiment that would test your prediction.

_____
_____
_____
_____

**Inquiry**

Think of your own questions that you might like to test. How are stomata related to water loss in the leaf?

My Question Is:

_____
_____

How I Can Test It:

_____
_____
_____
_____

My Results Are:

_____
_____
_____

Unit A · Characteristics of Living Things    Use with textbook page A67

Name _____ Date _____

**Alternative Explore**
Lesson 6

# Compare Leaves of Seed Plants

## Procedure

1. Obtain three different leaves.
2. Place one of the leaves between two pieces of paper. Rub a crayon over the top sheet of paper above the leaf.
3. Label the rubbing with the name of the plant.
4. Repeat the procedure with the other two leaves.
5. Draw the vein patterns that you can see in the rubbings.

**Materials**

- leaves from three different kinds of seed plants
- crayons
- paper

## Drawing Conclusions

1. Which plant has the largest leaf?
   _____

2. Which plant has the thickest leaf?
   _____

3. Are the vein patterns the same in all three leaves? Describe the patterns.
   _____
   _____

28    Unit A · Characteristics of Living Things    Use with TE textbook page A67

Name_____  Date_____

**Inquiry Skill Builder**
**Lesson 6**

# Observe

## Flowering Plants

In this activity you will observe flowering plants in order to classify them. That is, you will examine several plants and try to determine whether each is a monocot or a dicot. As you examine each plant sample, refer to the chart below to help you classify the sample.

**Materials**
- sample leaves and flowers from various angiosperms

## Procedure

1. **Observe** Get together with a few of your classmates and go on a leaf-and-flower-collecting field trip. (Make sure to avoid poison ivy, poison oak, and poison sumac leaves. Your teacher can tell you how to spot them.)

2. **Observe** Find a number of different angiosperms. Try to get a sample of a leaf and flower from each plant. If you can't get a flower, a leaf will do.

3. **Interpret Data** Look at the chart of Characteristics of Monocots and Dicots. It will give you clues on how to tell if the sample leaves and flowers you chose are monocots or dicots.

| Characteristics of Monocots and Dicots | | |
|---|---|---|
| **Characteristics** | **Monocots** | **Dicots** |
| Cotyledons | One | Two |
| Leaf veins | Parallel | Branched |
| Flower parts | Multiples of three | Multiples of four or five |
| Vascular system | Scattered in bundles | In rings |

Unit A · Characteristics of Living Things    Use with textbook page A73

Name_____ Date_____

**Inquiry Skill Builder**
**Lesson 6**

## Drawing Conclusions

1. **Observe** Examine the plant parts you have chosen. For each sample leaf, describe how the leaf veins look. For each sample flower, tell how many parts each flower has. Record your answers.

   _____
   _____
   _____
   _____

2. **Classify** Mount the leaves and flowers on a heavy sheet of cardboard, and indicate whether each came from a monocot or a dicot.

Name_____ Date_____

# How Do Flowers Differ?

**Hypothesize** Are all flowers alike? If not, how are flowers different? How are they alike? What do you think plants use their flowers for?

Write a **Hypothesis:**

_____

_____

### Materials
- several large flowers from different plants
- hand lens
- forceps
- dropper
- toothpick

## Procedure: Design Your Own

1. Decide how you will compare the flowers you look at. You may choose to look for parts that they seem to have in common. Describe what the parts are and how they differ from plant to plant.

   _____

   _____

   _____

   _____

   _____

2. Begin by removing the outer leaflike parts. Examine them. On a separate sheet of paper, draw what they look like.

3. Remove the petals. Examine them. Draw what they look like.

4. **Observe** Examine the rest of the flower. Draw what you see.

5. **Communicate** Draw the parts you decided to compare in different flowers.

Name _____ Date _____

## Drawing Conclusions

1. **Communicate** What color is each flower? What do you think the job of the petals is? How would you design an experiment to find out?

   _____
   _____
   _____

2. **Infer** What do you think the flower parts you chose are for? Do you think the same parts of different flowers do the same kinds of jobs for their plants?

   _____
   _____
   _____

3. **FURTHER INQUIRY** **Infer** Why do you think a plant has flowers? Design an experiment to test your hypothesis. Try it and report your results.

   _____
   _____
   _____

### Inquiry

Think of your own questions that you might like to test. How do the inner parts of the flower help the plant to reproduce?

My Question Is:

_____

How I Can Test It:

_____
_____
_____

My Results Are:

_____

Name_____ Date_____

**Alternative Explore**
Lesson 7

# Design a Flower

### Procedure

1. With your partner, look at the pictures and samples. Discuss what each flower part might do.
2. Design your own flower. List the ideas you and your partner have.

_____

_____

_____

3. Draw your design in the space below.
4. Show your design to the class and explain why you designed the flower this way.

**Materials**
- samples and pictures of different flowers

### Drawing Conclusions

1. Does you flower have large petals? Explain your decision.

_____

2. What are the tall, thin structures in the middle of most flowers for? Did you include any in your model?

_____

_____

_____

Name_____ Date_____

**QUICK LAB**
FOR SCHOOL OR HOME
Lesson 7

# Inside a Seed

**Hypothesize** What does a seed do? Where does it store its food? How do different seeds compare? Write a **Hypothesis:**

_____
_____
_____
_____

**Materials**
- bean seed (such as a lima bean)
- corn seed
- hand lens
- water

**Procedure**

1. Soak the bean seed in water overnight.
2. **Observe** Carefully pull apart the two halves of the seed. Examine the halves with a hand lens. Draw what you see.

Name_____ Date_____

**Drawing Conclusions**

3. **Infer** Which part of the seed is the embryo?
   _____
   _____

4. On your drawing label the seed coat and the cotyledon where food is stored.
   _____

5. **Communicate** Compare a corn kernel with a lima bean. Describe how its parts are similar to or different from the lima bean.
   _____
   _____
   _____

6. **Classify** Which seed is a dicot? Which is a monocot? Explain how you know which is which.
   _____
   _____

7. **Going Further** Think of your own questions you might like to test. What type of vein structure would you expect for corn and bean leaves?

   My Question Is:
   _____
   _____

   How I Can Test It:
   _____
   _____

   My Results Are:
   _____
   _____
   _____

Unit A · Characteristics of Living Things   Use with textbook page A82

Name_____ Date_____

**Explore Activity**
Lesson 8

# What Are the Traits of Animals?

**Hypothesize** Are animals grouped by their visual characteristics? Test your ideas.

Write a **Hypothesis:**

_____

_____

**Materials**
- 25 pictures of animals
- 5 sheets of paper
- tape
- scissors

**Procedure** **BE CAREFUL!** Be careful using scissors.

1. Cut out 25 animal pictures from old magazines.

2. **Classify** Think about the kinds of things all of the animals you found need to survive. Then, think about the traits that enable these animals to fulfill those needs.

_____

_____

3. **Communicate** Write why you think these animals are classified as animals.

_____

_____

4. **Classify** Now that you have seen the animals' similarities, look at their differences. What traits would you use to classify the 25 animals you have into different groups? How many groups would you make?

## Drawing Conclusions

1. Which trait was used most often for grouping the pictures?

_____

_____

2. **Infer** What is the best method for grouping the animals?

_____

_____

Name_____ Date_____

**Explore Activity**
**Lesson 8**

3. **FURTHER INQUIRY** **Infer** Why do you think scientists all over the world use a single system for grouping organisms?

_____

_____

**Inquiry**

Think of your own question that you might like to test. How else could you classify animals? Might you classify them using their shapes and skin textures?

My Question Is:

_____

_____

How I Can Test It:

_____

_____

My Results Are:

_____

_____

Unit A · Characteristics of Living Things     Use with textbook page A93

Name_____ Date_____

# Animal Traits

**Procedure**

1. Choose two different kinds of animals.

2. Draw a Venn diagram comparing and contrasting the traits of both animals. In each circle, list the traits the two animals do not have in common. Where the circles overlap, list the traits the two animals share.

_____
_____

**Drawing Conclusions**

3. Compare diagrams with a partner. What traits do all four of your animals share?

_____
_____
_____
_____

Name_____ Date_____

**Inquiry Skill Builder**
Lesson 8

# Make a Model

## Model a Backbone

Vertebrates have an internal skeleton with a backbone. Skeletons are made of bones or cartilage that give the body its overall shape. In this activity, you will learn more about the structure of a backbone as you make a model.

**Materials**
- pasta wheels
- soft-candy circles
- craft sticks
- hard candy circles

 Do not eat anything in the lab.

## Procedure

1. Use pasta wheels, soft-candy circles, and a craft stick to make a model of a backbone.

2. Alternately string the pasta wheels and the soft-candy circles on the craft stick until the row of candy and pasta is about 10 cm long.

3. Fold each end of the craft stick so the pasta wheels and candy do not come off.

4. Slowly bend the model. Does it move easily?

   _____

5. How far can you bend the model?

   _____

6. Compare your backbone to the model.

   _____

Unit A · Characteristics of Living Things

# Inquiry Skill Builder
## Lesson 8

## Drawing Conclusions

1. Build a model using hard candy circles with the pasta wheels. Compare the two models. Which model allows for more flexibility?

_____
_____

Name_____ Date_____

**Explore Activity**
Lesson 9

# How Do Sow Bugs Adapt to Their Environment?

**Materials**
- 10 sow bugs
- tray
- paper towels
- water

**Hypothesize** Do animals such as sow bugs adapt to their environments? Test your ideas. Write a **Hypothesis:**
_____
_____

**Procedure** BE CAREFUL! Handle live animals with care. Wash your hands well when you finish this activity.

1. **Observe** Place a sow bug in the center of the tray, and observe it. What traits does it have that enable it to live in the soil and under decaying wood or leaves? Record your observations.
   _____
   _____

2. **Observe** Touch the sow bug. How does it react?
   _____
   _____

3. **Experiment** Place all of the sow bugs in the center of the tray. Do the animals tend to stay together?
   _____
   _____

4. **Experiment** Move the sow bugs to one end of the tray. Dampen three or four paper towels, and place them in the opposite end of the tray. Observe for several minutes. Record your observations. When the animals move, do they tend to move faster in the dry section or wet section of the box?
   _____
   _____
   _____

Unit A · Characteristics of Living Things    Use with textbook page A105

Name_____ Date_____

## Drawing Conclusions

1. **Infer** How do sow bugs protect themselves?
   _____
   _____

2. **Infer** Can the behavior of sow bugs when exposed to moisture be related to their survival? If so, how?
   _____
   _____

3. **FURTHER INQUIRY** **Experiment** Design an experiment to test the reactions of sow bugs to light. Try it and report your results.
   _____
   _____
   _____
   _____

### Inquiry

Think of your own question that you might test. How might the sow bugs react to a predator?

My Question Is:

_____
_____

How I Can Test It:

_____
_____

My Results Are:

_____
_____

42   Unit A · Characteristics of Living Things          Use with textbook page A105

Name_____ Date_____

**Alternative Explore**
**Lesson 9**

# Earthworms and Light

## Procedure

**BE CAREFUL!** Be careful handling live animals.
Wash your hands well when you finish this activity.

1. Fill the jar three-quarters full with moist soil. Place a tablespoon of cornmeal on top.
2. Now add the earthworms. Cover the top of the jar with foil.
3. After 24 hours, take off the foil. Write your observations of the earthworms.
   _____
4. Replace the foil. Cover the outside of the jar with the dark paper.
5. Wait another 24 hours. Take off the foil and paper. Where are the earthworms now?
   _____

### Materials

- 3 earthworms
- moist soil
- plastic jar
- foil
- dark paper
- cornmeal
- tablespoon

## Drawing Conclusions

1. Describe how the earthworms respond to light.
   _____
   _____

2. Why do the earthworms react this way to light?
   _____
   _____

Unit A · Characteristics of Living Things        Use with TE textbook page A105

Name _____ Date _____

# Find the New Breed

**QUICK LAB**
FOR SCHOOL OR HOME
Lesson 9

**Hypothesize** Do hybrids exhibit traits of their parents? Test your ideas. Write a **Hypothesis:**

_____

_____

**Materials**
- pictures of Persian, Himalayan, and Siamese cats

**Procedure**

1. **Observe** Look at the picture of the Siamese cat on page A113 of your textbook. What traits do you think it has been bred for?

   _____

   _____

2. **Observe** Look at the picture of the Persian cat on page A113. What traits do you think it has been bred for?

   _____

   _____

3. **Observe** Look at the picture of the Himalayan cat on page A113. What traits do you think it has been bred for?

   _____

   _____

Name_____ Date_____

**QUICK LAB**
FOR SCHOOL OR HOME
Lesson 9

## Drawing Conclusions

4. **Infer** Which cat do you think is the new breed? Explain your answer.

   _____
   _____

5. **Going Further: Predict** Choose two different but related animals as possible hybrid parents. What desirable traits might the offspring of these two animals have?

   _____
   _____
   _____
   _____

Unit A · Characteristics of Living Things     Use with textbook page A113

Name_____ Date_____

**Explore Activity**
Lesson 1

# What Do Living Things Need to Survive?

**Hypothesize** How do living things interact with each other and their environment? What do living things need in order to survive? How would you design a special environment to test your ideas?

Write a **Hypothesis:**

_____
_____
_____
_____

### Materials

- wide-mouthed, clear container with lid
- washed gravel
- pond water or aged tap water
- water plants
- water snails
- soil
- small rocks
- grass seed and small plants
- earthworms, land snails, sow bugs, or other small land animals that eat plants

**BE CAREFUL!** Handle animals and plants gently.

## Procedure: Design Your Own

1. For a water environment, add thoroughly washed sand or gravel to the jar. Fill the jar with water. Add a few floating plants, rooted plants with floating leaves, and submerged plants. Add water snails.

2. For a land environment, place a layer of gravel on the bottom of the jar. Cover the gravel layer with a layer of moistened soil. Add plants, and plant grass seeds. Add earthworms, sow bugs, and snails.

3. Place each jar in a lighted area but not in direct sunlight.

4. Cover each jar with its own lid or with a piece of plastic wrap. Record the number and types of living things you used.

_____
_____
_____

46  Unit B·Living Things and Their Environment    Use with textbook page B5

Name_____ Date_____

5. **Observe** Examine your jars every other day. Record your observations on another sheet of paper.

## Drawing Conclusions

1. **Infer** What are the nonliving parts of your system? What are the living parts of your system?

   _____
   _____

2. **Infer** What do the living things need to survive? How do you know?

   _____
   _____

3. **FURTHER INQUIRY** **Experiment** How could you design an environment that contains both land and water areas?

   _____
   _____

### Inquiry

Think of your own questions that you might like to test. How do changes in conditions in an environment affect the organisms in the environment?

My Question Is:

_____
_____

How I Can Test It:

_____
_____

My Results Are:

_____
_____
_____

Name_____ Date_____

Lesson 1

# Plant Needs

In 1699, a scientist named John Woodward did an experiment in plant growth. He weighed four plants and put each in a flask. He added different water samples to each one—rainwater, muddy river water, drain water, and tap water with partly rotted leaves. He weighed the plants to see which had increased its weight the most.

### Materials

- 4 water mint plants
- 4 flasks
- marking pen
- 4 different water samples
- scale
- ruler

## Procedure

1. With your group, decide on four different water samples you will use to do an experiment like Woodward's.

2. With your group, decide how you will carry out your experiment. Write the steps your group plans for the experiment.

   _____
   _____
   _____
   _____

3. Show the plan to your teacher for approval. Then begin to carry out your plan. Use a separate sheet of paper to record your data and observations.

## Drawing Conclusions

1. What four types of water samples did your group decide to test?

   _____
   _____

2. How did you decide to measure plant growth?

   _____

3. Did any plant grow better than others? If so, which one?

   _____
   _____

Name_____ Date_____

# Changing the Environment

**Hypothesize** How do animals change their environment? Write a hypothesis.

_____

_____

## Procedure

1. Select a wild animal that you find interesting. It can be as small as an insect or as large as a whale.

   _____

2. Do research to find where this animal lives and what it does to survive in its environment.

3. Draw or find a picture of the animal you selected.

Name_____ Date_____

**QUICK LAB**
FOR SCHOOL OR HOME
Lesson 1

4. **Communicate** How does the animal you selected change the environment where it lives? What living and nonliving things are affected by these changes? Write your answers.

_____
_____
_____

5. **Infer** How do the living things you listed above adapt to the changes in their environment?

_____
_____

6. **Going Further** How do animals change their environment? Write and conduct an experiment.

My Hypothesis Is:

_____

My Experiment Is:

_____
_____

My Results Are:

_____

50 | Unit B·Living Things and Their Environment | Use with textbook page B13

Name_____ Date_____

**Explore Activity**
Lesson 2

# How Do Populations Interact?

**Hypothesize** How can changes in a population lead to changes to the ecosystem in which the population lives? What kinds of changes might these be? How might you test your ideas? Write a **Hypothesis:**

_____
_____

**Materials**
- tape
- string
- population cards, p.53

## Procedure

1. Cut out the cards representing the plants and animals in the ecosystem.
2. Label the top of your paper *Sunlight*.
3. Place the plant cards on the paper, and link each to the sunlight with tape and string.
4. Link each plant-eating animal to a plant card. Link each meat-eating animal to its food source. Only two animals can be attached to a food source. Record the links you have made.

_____
_____
_____

5. Fire destroys half the plants. Remove four plant cards. Rearrange the animal cards. Remove animal cards if more than two animals link to any one food source. Record the changes you have made.

## Drawing Conclusions

1. **Observe** What has happened to the plant eaters as a result of the fire? To the animal eaters?

_____
_____
_____

Name_____ Date_____

**Explore Activity**
**Lesson 2**

2. **Infer** Half of the plants that were lost in the fire grow back again. What happens to the animal populations?

   _____
   _____

3. **Experiment** Try adding or removing plant or animal cards. What happens to the rest of the populations?

   _____
   _____

4. **FURTHER INQUIRY** **Predict** If plants or prey become scarce, their predators may move to a new area. What will happen to the ecosystem the predators move into?

   _____
   _____
   _____

**Inquiry**

Think of your own questions you might like to test. How would an ecosystem change if some meat-eating animals were removed from it?

My Question Is:

_____
_____

How I Can Test It:

_____
_____

My Results Are:

_____
_____
_____

Name_____ Date_____

**Bison**
Food: prairie plants

**Prairie Plants**
Food: made from water, carbon dioxide, and sunlight

**Field Sparrow**
Food: prairie plants

**Prairie Plants**
Food: made from water, carbon dioxide, and sunlight

**Lizard**
Food: insects

**Prairie Plants**
Food: made from water, carbon dioxide, and sunlight

**Pronghorn Antelope**
Food: prairie plants

**Racer (snake)**
Food: lizards, mice, insects

**Meadowlark**
Food: crickets, grasshoppers

**Coyote**
Food: rabbits, ground squirrels, meadow mice, other rodents

**Prairie Plants**
Food: made from water, carbon dioxide, and sunlight

**Prairie Plants**
Food: made from water, carbon dioxide, and sunlight

**Bullsnake**
Food: mice, rabbits, ground squirrels, birds and eggs

**Field Cricket**
Food: prairie plants, other insects

**Ground Squirrel**
Food: prairie plants

**Red-Tailed Hawk**
Food: ground squirrels, mice, rabbits, snakes, lizards, small birds

**Badger**
Food: ground squirrels, rabbits, mice, lizards

**Prairie Plants**
Food: made from water, carbon dioxide, and sunlight

**Grasshopper**
Food: prairie plants

**Prairie Plants**
Food: made from water, carbon dioxide, and sunlight

**Cottontail Rabbit**
Food: prairie plants

**Prairie Plants**
Food: made from water, carbon dioxide, and sunlight

**Meadow Mouse**
Food: prairie plants

Unit B·Living Things and Their Environment    Use with textbook page B17

**Name** _____ **Date** _____

**Alternative Explore**
Lesson 2

# Food Chain Model

## Procedure

1. Cut strips of construction paper. The strips will represent the links in a food chain that includes the Sun, plants, a plant-eater, and a meat-eater. Label the links with these four things.

2. Lay out the strips of paper in the order that they should appear in a food chain. Choose specific organisms for each step in your food chain. Use reference materials to select your organisms. List the organisms you chose.

   _____
   _____
   _____

3. Write the name of each organism on the correct piece of paper.
4. Use tape to make the paper strips into links that form a chain.
5. Share your chain with other students. Look for ways your chains are similar and ways they are different.

### Materials
- construction paper
- tape
- scissors
- reference materials

## Drawing Conclusions

1. In what ways were all the food chains similar?

   _____

2. In what ways did the food chains differ from one another? Give an example.

   _____
   _____
   _____

Name_____ Date_____

# Getting Food

**Quick Lab** — FOR SCHOOL OR HOME — Lesson 2

**Hypothesize** What living things are in your community? Which are producers? Which are consumers? Write a Hypothesis:

_____
_____

**Materials**
- camera (optional)
- collecting net (optional)

## Procedure

1. Take a walk outdoors around your home or school. Choose a community to study. Make a list of the living things you see. Don't include people or domestic animals like dogs, cats, and farm animals. You may want to take photos to complete your observations. Use illustrations to complete step 2.

_____
_____
_____
_____

2. **Classify** Set the organisms into two groups—those that can make their own food (producers) on the left and those that cannot (consumers) on the right.

| Producers | Consumers |
|---|---|
|  |  |
|  |  |
|  |  |
|  |  |
|  |  |
|  |  |
|  |  |
|  |  |

Unit B·Living Things and Their Environment — Use with textbook page B19

Name _____ Date _____

**Inquiry Skill Builder**
**Lesson 5**

# Use Numbers

## How Can You Find Probability Using a Punnett Square?

Eyelashes can be long or short. Long eyelashes are the dominant trait. Short eyelashes are the recessive trait. To represent these traits, you can use the letters:

- L (the factor for long eyelashes, dominant)
- l (the factor for short eyelashes, recessive)

A person with long eyelashes may have either two of the same factors (LL) or two different factors (Ll). A person with short eyelashes has two of the same factors (ll).

## Procedure

1. **Make a Model** Two parents both have long eyelashes. Each parent has two different factors for eyelash length. Draw a Punnett square to show what factors for eyelash length a child might inherit from the parents.

| Parent Factors | | |
|---|---|---|
| | | |
| | | |

2. How many different Punnett squares can you draw to show one parent with long eyelashes and one with short? Draw your Punnett squares below.

Name_____ Date_____

## Explore Activity
### Lesson 3

# What Controls the Growth of Populations?

**Hypothesize** What kinds of things do organisms need in their environment in order to survive? What happens when these things are limited or unavailable? Test your ideas. Write a **Hypothesis**:

_____
_____

### Materials
- 4 small, clean milk cartons with the tops removed
- 40 pinto bean seeds that have been soaked overnight
- soil
- water

## Procedure

1. Label the cartons 1 to 4. Fill cartons 1 and 2 with dry potting soil. Fill cartons 3 and 4 with moistened potting soil. Fill the cartons to within 2 cm of the top.

2. Plant ten seeds in each carton, and cover the seeds with 0.5 cm of soil.

3. **Use Variables** Place cartons 1 and 3 in a well-lighted area. Place cartons 2 and 4 in a dark place. Label the cartons to show if they are wet or dry and in the light or in the dark.

4. **Observe** Examine the cartons each day for four days. Keep the soil moist in cartons 3 and 4. Record your observations.

_____
_____

5. Observe the plants for two weeks after they sprout. Continue to keep the soil moist in cartons 3 and 4, and record your observations.

_____
_____
_____

Name_____ Date_____

**Explore Activity**
**Lesson 3**

## Drawing Conclusions

1. **Communicate** How many seeds sprouted in each carton?
   _____
   _____

2. **Observe** After two weeks how many plants in each carton were still living?
   _____
   _____

3. Why did some seeds sprout and then die?
   _____
   _____

4. **FURTHER INQUIRY** **Infer** Use your observations to explain what is needed for seeds to sprout and what is needed for bean plants to grow. Use evidence to support your explanation.
   _____
   _____
   _____

### Inquiry

Think of your own questions you might like to test. Do plants need anything other than water and sunlight to thrive?

My Question Is:
_____

How I Can Test It:
_____
_____
_____

My Results Are:
_____

Name_____ Date_____

**Alternative Explore**
Lesson 3

# Effect of Water and Light on Plants

**Materials**
- 4 similar plants
- labels
- water

## Procedures

1. Label the plants 1, 2, 3, and 4.
2. Put plants 1 and 2 in a sunny location. Water plant 1 each day. Do not water plant 2.
3. Put plant 3 in a sunny location. Put plant 4 in a dark location. Give both plants the same amount of water.
4. Continue the experiment for two weeks. Make a table to record each day's observations on a separate sheet of paper.

## Drawing Conclusions

1. How did the growth of plants 1 and 2 compare? Explain any differences you saw.

   _____
   _____
   _____

2. How did the growth of plants 3 and 4 compare? Explain any differences you saw.

   _____
   _____
   _____

3. What can you conclude about the kinds of conditions that plants need to grow?

   _____
   _____

# Use Variables

**Inquiry Skill Builder — Lesson 3**

## Vanishing Bald Eagles

The table below shows the average number of bald eagle eggs that hatched in the wild during a 16-year period. It also shows the level of an insecticide in bald eagle eggs during the same period. What is the relationship between these two variables?

**Materials**
- ruler

Variables are things that can change. In order to determine what caused the results of an experiment, you need to change one variable at a time. The variable that is changed is called the *independent variable*. A *dependent variable* is one that changes because of the independent variable.

### BALD EAGLE EGG-HATCHING DATA

| Year | 1966 | 1967 | 1968 | 1969 | 1970 | 1971 | 1972* | 1973 | 1974 | 1975 | 1976 | 1977 | 1978 | 1979 | 1980 | 1981 |
|---|---|---|---|---|---|---|---|---|---|---|---|---|---|---|---|---|
| Average number of young hatched (per nest) | 1.28 | 0.75 | 0.87 | 0.82 | 0.50 | 0.55 | 0.60 | 0.70 | 0.60 | 0.81 | 0.90 | 0.93 | 0.91 | 0.98 | 1.02 | 1.27 |
| Insecticide in eggs (parts per million) | 42 | 68 | 125 | 119 | 122 | 108 | 82 | 74 | 68 | 59 | 32 | 12 | 13 | 14 | 13 | 13 |

*pesticide banned

## Procedure

1. **Infer** What is the independent variable in the study? What is the dependent variable in the study?

   _____

   _____

2. **Communicate** On a separate piece of paper, make a line graph showing the average number of young that hatched. Make another line graph showing the amount of insecticide in eggs.

Name_____ Date_____

**Inquiry Skill Builder**
**Lesson 3**

## Drawing Conclusions

1. **Use Variables** Based on the graphs, what appears to be the relationship between the amount of insecticide in eggs and the number of young hatched?

   _____
   _____
   _____
   _____

2. **Hypothesize** Suggest a reason for the relationship.

   _____
   _____
   _____

Name_____ Date_____

# What Is the Water Cycle?

**Hypothesize** How can we, and all living things, keep using water every day and not use it all up? How would you experiment to test your ideas?

Write a **Hypothesis:**

_____
_____
_____

### Materials
- plastic food container with clear cover
- small bowl or cup filled with water
- small tray filled with dry soil
- paper towel
- 100-W lamp (if available)

## Procedure

1. Place the dry paper towel, the dry soil, and the bowl of water in the plastic container. Close the container with the lid.

2. **Observe** Place the container under a lamp or in direct sunlight. Observe every ten minutes for a class period. Record your observations.

_____
_____

3. Observe the container on the second day. Record your observations.

_____
_____

## Drawing Conclusions

1. What did you observe the first day? What did you observe the second day?

_____
_____

Name_____ Date_____

**Explore Activity Lesson 4**

2. **Infer** What was the source of the water? What was the source of the energy that caused changes in the container?

_____
_____
_____

3. What happened to the water?

_____
_____
_____

4. **FURTHER INQUIRY** **Infer** How did the water move?

_____
_____
_____

## Inquiry

Think of your own questions you might like to test. What happens to the salt in ocean water when the water evaporates?

My Question Is:

_____

How I Can Test It:

_____
_____
_____
_____
_____

My Results Are:

_____
_____

Name_____ Date_____

# Recycling Water

**BE CAREFUL!** Be careful handling the hot plate and teakettle.

## Procedure

1. Put water in the teakettle and place the teakettle on the hot plate.
2. Heat the water until you see steam coming out of the teakettle.
3. Place several ice cubes on the cookie sheet.
4. Use the oven mitts to pick up the cookie sheet and hold it over the steam coming from the kettle.
5. Look at the underside of the cookie sheet. Record your observations.

_____

_____

### Materials

- hot plate
- cookie sheet
- water
- teakettle
- ice cubes
- oven mitts

## Drawing Conclusions

1. What happened to the water in the teakettle?

_____

2. What happened to the steam when it hit the cold cookie sheet?

_____

3. What would happen if you shook the cookie sheet?

_____

Name_____ Date_____

# Soil Sample

**Hypothesize** How do nutrients get recycled in nature?
Write a **Hypothesis:**
_____
_____

**Materials**
- empty can

## Procedure

**BE CAREFUL!** Do not touch the sharp edges of the can.

1. Go to a wooded area in a park or other location near your school. Find a patch of soft, moist soil.

2. Press a can, open side down, into the soil to get a core sample. You might have to gently rotate the can so it cuts into the soil.

3. **Observe** Carefully remove the core so it stays in one piece. Describe and draw the core.

Name_____ Date_____

## Drawing Conclusions

4. **Infer** From top to bottom, what kind of matter does the core hold? In what order did the layers form?

   _____
   _____
   _____
   _____

5. **Infer** Which layer holds the most available nutrients? Explain.

   _____
   _____
   _____

6. **Going Further** How do worms help return nutrients to the soil? Write and conduct an experiment.

   My Hypothesis Is:

   _____

   My Experiment Is:

   _____
   _____

   My Results Are:

   _____
   _____

Name_____ Date_____

## Explore Activity
### Lesson 5

# Why Is Soil Important?

**Hypothesize** Why is the soil in one kind of ecosystem different from the soil in another kind of ecosystem? What determines what the soil is like? Write a **Hypothesis:**

_____
_____
_____

**Materials**
- washed sand
- soil
- hydrogen peroxide
- 2 plastic cups
- 2 plastic spoons
- dropper
- goggles
- apron

## Procedure

**BE CAREFUL!** Wear goggles and an apron.

1. Place 1 tsp. of washed sand in a plastic cup.

2. **Observe** Using the dropper, add hydrogen peroxide to the sand, drop by drop. Count each drop. Bubbles will form as the hydrogen peroxide breaks down any decayed matter.

3. **Communicate** Record the number of drops you add until the bubbles stop forming.

   _____
   _____

4. **Experiment** Repeat steps 1–3 using the soil.

## Drawing Conclusions

1. Which sample—soil or sand—gave off more bubbles?

   _____
   _____

Name_____ Date_____

**Explore Activity — Lesson 5**

2. **Infer** Why was the sand used?

   _____
   _____
   _____

3. **Infer** Decayed materials in soil release their nutrients to form humus. The amount of humus in soil depends on the rate of decay and the rate at which plants absorb the nutrients. Which sample had more humus?

   _____
   _____

4. **FURTHER INQUIRY** **Infer** Use your observations to identify in which sample you could grow larger, healthier plants. Give evidence to support your answer.

   _____
   _____
   _____
   _____

**Inquiry**

Think of your own questions you might like to test. How much humus do other soil samples have?

My Question Is:

_____

How I Can Test It:

_____
_____

My Results Are:

_____
_____

Unit B · Living Things and Their Environment — Use with textbook page B63

Name_____ Date_____

# Testing Soil pH

## Procedure

1. Mix a spoonful of washed sand and a little distilled water in a petri dish.

2. Test the mixture with pH paper. What was the pH of the sand mixture?

   _____

   _____

3. Test the other soil sample(s) in the same way. Record your results in the table.

**Materials**
- washed sand
- distilled water
- pH test paper
- compost, potting soil, or garden soil
- petri dishes
- spoons

| Sample | pH |
|---|---|
|  |  |
|  |  |
|  |  |
|  |  |

## Drawing Conclusions

1. Which sample had the highest pH?

   _____

2. Which sample had the lowest pH?

   _____

3. A low pH means the sample is acidic and not as nutrient rich as soils with a higher pH. Which sample is the most nutrient-rich? Explain your answer.

   _____

   _____

   _____

Unit B·Living Things and Their Environment   Use with TE textbook page B63

Name_____ Date_____

**QUICK LAB**
FOR SCHOOL OR HOME
Lesson 5

# Freshwater Communities

**Hypothesize** Do different organisms live in different locations in aquatic ecosystems? Write a **Hypothesis:**

_____

_____

## Procedure

1. Obtain from your teacher samples of pond, lake, or stream water taken at different locations. Use a different container for each sample. Record on the container the location each sample came from.

2. **Observe** Place a drop of water on a slide, and carefully place a coverslip over it. Use low and high power to examine the slide under a microscope.

3. **Communicate** Sketch what you see.

_____

_____

_____

### Materials

- dropper
- microscope slide
- coverslip
- microscope
- at least 3 samples of pond, lake, or stream water
- 3 or more plastic containers with lids

## Drawing Conclusions

4. **Interpret Data** What does this tell you about aquatic ecosystems?

_____

Name_____ Date_____

**QUICK LAB**
FOR SCHOOL OR HOME
Lesson 5

5. **Going Further** How do organisms in a different aquatic environment compare to those you examined in the activity? Design and conduct an experiment.

My Hypothesis Is:

_____
_____
_____

My Experiment Is:

_____
_____
_____

My Results Are:

_____
_____

Name_____ Date_____

# How Do Ecosystems Change?

**Hypothesize** How can ecosystems change? How might an abandoned farm change? Test your ideas. Write a **Hypothesis:**

_____
_____

## Procedure

1. **Observe** Examine the drawing.
2. **Communicate** Describe what you see.

   _____
   _____
   _____
   _____

## Drawing Conclusions

1. **Infer** What happened to this farm after the owner left and moved to the city?

   _____
   _____
   _____

2. **Infer** Think about how this farm might have looked ten years ago. What kind of plants lived there then?

   _____
   _____

Name_____ Date_____

**Explore Activity**
**Lesson 6**

3. **Interpret Data** How can one ecosystem be changed into another?

   _____

4. Compare what you think will happen to the abandoned farm with what happened at Mount Saint Helens. In what ways would the changes in the ecosystems be similar? In what ways would they be different?

   _____
   _____
   _____
   _____

5. **FURTHER INQUIRY** **Predict** Think of another ecosystem that might be changed by nature. Think of an ecosystem that might be changed by humans. Describe how such ecosystems might continue to change over time.

   _____
   _____
   _____
   _____
   _____

**Inquiry**

Think of your own questions you might like to test. How has another ecosystem been changed by nature?

My Question Is:

_____

How I Can Test It:

_____

My Results Are:

_____
_____
_____

Name_____  Date_____

**Alternative Explore** — Lesson 6

# From Pond to Forest

## Procedure

1. Draw a side view of a pond showing plants growing at the edges in the space below. Use extra paper if necessary.

**Materials**
- reference materials

2. Do research to find out what happens to a pond over time as the plants grow.
3. Draw three more pictures of your pond, showing how it could change into a new ecosystem.

## Drawing Conclusions

1. What plants grow into the pond at the beginning of this process?

   _____

2. Where do the nutrients for more plants in the pond come from?

   _____

   _____

3. What do you think would be the stage after the woodlands? Explain.

   _____

   _____

Name_____ Date_____

# Predicting Succession

**Hypothesize** In what areas where you live do you think ecological succession may be taking place? Write a **Hypothesis:**

_____
_____

## Procedure

1. **Observe** Identify an area near you where you think ecological succession is taking place.

   _____
   _____

2. **Communicate** Describe the area. List the evidence you have that indicates ecological succession is taking place.

   _____
   _____
   _____
   _____
   _____
   _____

## Drawing Conclusions

3. **Infer** Do you think the succession will be primary or secondary? Explain.

   _____
   _____
   _____
   _____
   _____
   _____

Name_____ Date_____

4. **Predict** In what order do you think new species will colonize the area? Explain the reasons for your predictions.

   _____
   _____
   _____
   _____

5. **Communicate** Describe the climax community that you think will eventually live in the area. Give reasons for your conclusion.

   _____
   _____
   _____
   _____
   _____
   _____

6. **Going Further** Is ecological succession taking place in other areas near you? Write and conduct an experiment.

   My Hypothesis Is:

   _____
   _____

   My Experiment Is:

   _____
   _____

   My Results Are:

   _____
   _____
   _____

Name_____ Date_____

# Infer

**Inquiry Skill Builder** — Lesson 6

## Comparing Ecosystems in Volcanic Areas

In this activity you will collect data and infer about the ecosystems of two volcanic areas.

Data are different kinds of facts. They might include observations, measurements, calculations, and other kinds of information. Scientists collect data about an event to better understand what caused it, what it will cause, and how it will affect other events.

What do these data tell the scientist? The scientist first organizes the data in some way—perhaps a table, chart, or graph. The scientist then studies the organized data and makes inferences. To infer means to form an idea from facts or observations. In this case you will infer about which plants will return to a volcanic area.

**Materials**
- research books
- Internet

## Procedure

1. Collect data on two volcanic areas, such as Mount Saint Helens and the Soufriere Hills volcano on the island of Montserrat or the active volcanoes of Hawaii. Organize the data on a separate sheet of paper.

2. **Communicate** Describe the sequence of events that has taken place.

_____
_____
_____
_____
_____

3. **Interpret Data** Draw a conclusion about why certain plants return when they do.

_____
_____
_____
_____
_____
_____

Unit B·Living Things and Their Environment — Use with textbook page B87

Name_____ Date_____

# Inquiry Skill Builder
## Lesson 6

## Drawing Conclusions

1. In what ways is succession in the two areas alike? In what ways is it different?

   _____
   _____
   _____
   _____

2. **Infer** Why is the succession in these two areas similar or different?

   _____
   _____
   _____
   _____
   _____
   _____

3. **Infer** What abiotic factors must you consider when drawing conclusions? What biotic factors must you consider?

   _____
   _____
   _____
   _____
   _____
   _____
   _____
   _____

Name_____ Date_____

**Explore Activity**
Lesson 1

# What Makes the Crust Move?

**Hypothesize** What kind of motion causes an earthquake? Does it always cause destruction? Can it result in anything else? Test your ideas. Write a **Hypothesis:**

_____
_____
_____

**Materials**
- 4–6 matching books (optional)
- layers of clay or modeling compound (optional)
- plastic knife (for use with clay)
- cubes
- wax paper

## Procedure: Design Your Own

1. **Make a Model** Work with a partner to model layers of rock. You may use books, clay, or other materials to represent rock layers. Build your model on wax paper. Include a "crack" down through the layers. Stack cubes on the top of the model to represent buildings and other surface features.

2. **Experiment** Find as many ways of moving the model as you can to show how the crust may move during an earthquake. What happens to the surface features as you move the model each way? Draw and describe each.

_____
_____
_____

3. **Experiment** How can you show movement without causing any visible effect on the surface features?

_____
_____

## Drawing Conclusions

1. How many different ways could you move your model? How were they different?

_____
_____

Unit C · Earth and Its Resources          Use with textbook page C5

Name_____ Date_____

**Explore Activity**
**Lesson 1**

2. **Communicate** How did each way you moved the model affect the surface features? How did each way change the positions of the layers? Explain.

_____
_____
_____
_____

3. **Communicate** How did you move the model without moving the surface features? Did the model change in any way? Explain.

_____
_____

4. **FURTHER INQUIRY** **Experiment** How can you use your model to show how a mountain might rise up high above sea level? Explain and demonstrate.

_____
_____

**Inquiry**

Think of your own questions that you might like to test. What other effects can you demonstrate with the clay model? (Sometimes enough pressure is created when plates collide that part of a plate melts.)

My Question Is:

_____

How Can I Test It:

_____
_____

My Results Are:

_____
_____

Name_____ Date_____

Lesson 1

# Motion of the Crust

## Procedure

**Materials**
- newspapers
- scissors or ruler

1. Place a stack of newspapers 1 cm thick on a table.

2. Place your hands on the edges of the stack, and push toward the center of the paper. Observe how the paper moves. Record your observations.

   _____
   _____

3. Cut several thicknesses of newspaper into two pieces, or tear them using a ruler as a straightedge. Then place the edges together again. Try to line up the lines of type. Move the two stacks in opposite directions. Record your observations.

   _____
   _____

4. Place the cut halves of paper on the desk, a few centimeters apart. Place your hands on the paper and quickly slide the two halves toward each other. Observe what happens. Record your observations.

   _____
   _____

## Drawing Conclusions

Which way of moving the papers could be a model for how mountains are formed?

_____

Unit C · Earth and Its Resources     Use with textbook page C5

Name_____ Date_____

**FOR SCHOOL OR HOME**
**Lesson 1**

# Model of Earth

**Hypothesize** Can materials with different properties be used to make a solid Earth? Write a **Hypothesis:**

_____

_____

## Procedure

**BE CAREFUL!** Students who are allergic to peanuts should not do this activity!

1. **Infer** You will use four materials to make a model of Earth on wax paper. Each material is one of Earth's layers. Read step 2. Decide which material represents which layer. Decide how thick each layer needs to be.

2. **Make a Model** Wash your hands. Cover a hazelnut with a layer of peanut butter. Put the covered nut in a plastic bag of mashed banana so that the banana covers it completely. Roll the result into graham cracker crumbs on wax paper.

### Materials
- mashed ripe banana (in a plastic bag)
- peanut butter
- hazelnut
- graham cracker crumbs (in a plastic bag)
- wax paper

## Drawing Conclusions

3. How does each material represent a different layer?

_____

_____

_____

_____

82     Unit C · Earth and Its Resources     Use with textbook page C7

Name_____ Date_____

4. How thick did you decide to make each layer? Explain your reasoning.

   _____
   _____
   _____
   _____

5. **Going Further** Make a model of the plates in Earth's crust. Mix dirt and water together to make mud. Pour the mud on to a cookie sheet. Place the cookie sheet in the sun for several days to dry the mud. When the mud is completely dry, press on the outer edges. What happened to the dry mud when you pressed on it? How can this be related to Earth's crust?

   _____
   _____
   _____
   _____

   Can you relate this model to Earth's continents?

   _____
   _____
   _____

Name _____ Date _____

# How Does Steepness of Slope Affect Stream Erosion?

**Explore Activity**
**Lesson 2**

**Hypothesize** Water flowing in rivers and streams can pick up and carry sediments like silt, sand, and gravel. How does steepness of slope affect how fast a stream flows? How does this affect the size and amount of material rivers can carry? Write a possible explanation, or hypothesis to answer this question:

_____
_____

**Materials**
- long cake pan
- mixture of sand, coarse gravel, pebbles
- sprinkle bottle
- water
- books or wood blocks

## Procedure: Design Your Own

1. **Make a Model** Make a model to test your hypothesis. What materials will you need? What will you do? Which factors will you control? What factor will you manipulate?

_____
_____
_____
_____
_____

2. **Experiment** Set up your model and carry out your experiment. Make a chart to organize your data.

_____
_____

## Drawing Conclusions

1. **Interpret Data** Does the data you collected support your hypothesis? Explain. Compare your data with others.

_____
_____

Unit C · Earth and Its Resources    Use with textbook page C19

Name _____ Date _____

**Explore Activity Lesson 2**

2. **Communicate** Write a short newspaper article explaining how steepness of slope affects stream erosion.

_____

_____

3. **FURTHER INQUIRY** **Experiment** How does the volume of water affect stream erosion? Propose a hypothesis and experiment to test it.

_____

_____

**Inquiry**

Think of your own questions you might like to test.

My Question Is:

_____

How Can I Test It:

_____

My Results Are:

_____

Unit C · Earth and Its Resources — Use with textbook page C19

Name_____ Date_____

# Erosion on Slopes

**Procedure**

1. Make two models of slopes with the same amount of sand and gravel on each. Set one pan on a slight hill and the other on a steeper hill.

2. Pour the same amount of water over each hill at the same speed and distance from the hill.

3. Measure the amount of sand and gravel in the bottom of the pan to check the amount of erosion. Observe what happens. Record your observations below.

_____

_____

**Materials**

- modeling clay
- two 9-inch square cake pans
- sand
- gravel
- sprinkling can
- measuring cup

Name_____ Date_____

# Erosion Challenge

## Procedure

1. Cover your workspace with newspaper. Gather different samples of materials from your teacher, such as soil, sand, gravel, and plastic plants, as well as a foil pan.

2. Working with a partner, design a "hill" that is resistant to erosion. You may choose which materials to use and how to place them. Think about the effects of slope on the rate of water flow. Build your hill inside the foil pan. Sketch and describe your hill below.

**Materials**
- soil
- sand
- gravel
- plastic plants
- foil pans
- sprinkling cans
- newspaper

Unit C · Earth and Its Resources    Use with textbook page C23    87

Name_____ Date_____

**QUICK LAB FOR SCHOOL OR HOME**
**Lesson 2**

3. After you have built your hill, use a sprinkling can to create a rain shower over each group's hill. Compare your results with your classmates.

4. **Use Variables** Repeat step 2 to see how you could improve your hill's rate of erosion. Draw your improved hill below.

5. **Observe** How did the slope of the hill affect the rate of runoff?
   _____

6. **Observe** What other factors affected the rate of runoff?
   _____
   _____

7. **Infer** How could a farmer use this information to help prevent erosion?
   _____

Name_____ Date_____

# Explore Activity
**Lesson 3**

# How Can You Identify a Mineral?

**Hypothesize** How do you think people can tell minerals apart? Write a **Hypothesis**:

_____
_____
_____

### Materials
- mineral samples
- clear tape
- red marker
- copper penny or wire
- porcelain tile
- hand lens
- nail

## Procedure

1. **Communicate** Use tape and a marker to label each sample with a number. On another sheet of paper, make a table with the column headings shown. Fill in the numbers under "Mineral" to match your samples.

   Color = color of surface

   Porcelain Plate Test = the color you see when you rub the sample gently on porcelain

   Shiny Like a Metal = reflects light like a metal, such as aluminum foil or metal coins

   Scratch (Hardness) = Does it scratch copper? A nail?

   Other: Is it very dense? (Is a small piece heavy?) Has it got flat surfaces?

|   | Mineral | Color | Shiny like a Metal (Yes/No) | Porcelain Plate Test | Scratch (Hardness) | Other |
|---|---------|-------|------------------------------|----------------------|---------------------|-------|
| 1. |        |       |                              |                      |                     |       |
| 2. |        |       |                              |                      |                     |       |

2. **Observe** Use the table shown as a guide to collect data on each sample. Fill in the data in your table. Turn to the table on page 91 for more ideas to fill in "Other."

Unit C · Earth and Its Resources     Use with textbook page C31

Name _____ Date _____

**Explore Activity**
**Lesson 3**

## Drawing Conclusions

1. Use your data and the table on the third page of this activity to identify your samples. Were you sure of all your samples? Explain.

   _____
   _____
   _____

2. Which observations were most helpful? Explain.

   _____
   _____

3. **FURTHER INQUIRY** Experiment How could you make a better scratch (hardness) test?

   _____
   _____

**Inquiry**

Think of your own questions that you might like to test. Can minerals scratch each other?

My Question Is:

_____
_____

How I Can Test It:

_____
_____

My Results Are:

_____

Name_____ Date_____

| Mineral | Color(s) | Luster (Shiny as Metals) | Porcelain Plate Test (Streak) | Cleavage (Number) | Hardness (Tools Scratched by) | Density (Compared with Water) |
|---|---|---|---|---|---|---|
| | | **PROPERTIES OF MINERALS** | | | | |
| Gypsum | colorless, gray, white, brown | no | white | varies | 2 (all six tools) | 2.3 |
| Quartz | colorless, various colors | no | none | no | 7 (streak plate) | 2.6 |
| Pyrite | brassy, yellow | yes | greenish black | no | 6 (steel file, streak plate) | 5.0 |
| Calcite | colorless, white, pale blue | no | colorless, white | yes—3 | 3 (all but fingernail) | 2.7 |
| Galena | steel gray | yes | gray to black | yes—3 (cubes) | 2.5 (all but fingernail) | 7.5 |
| Feldspar | gray, green yellow, white | no | colorless | yes—2 | 6 (steel file, streak plate) | 2.5 |
| Mica | colorless, silvery, black | no | white | yes—1 (thin sheets) | 2—3 (all but fingernail) | 3.0 |
| Hornblende | green to black | no | gray to white | yes—2 | 5–6 (steel file, streak plate) | 3.4 |
| Bauxite | gray, red brown, white | no | gray | no | 1–3 (all but fingernail) | 2.0–2.5 |
| Hematite | black, gray, red-brown | yes | red or red-brown | no | 1–6 (all) | 5.3 |

Name_____ Date_____

# Classifying Minerals

## Procedure

1. Work with your group to develop ways to classify the mineral samples. You may classify them by color, feel, density, shape, or any other characteristics your group agrees on. List your group's suggestions.

   _____
   _____
   _____

**Materials**
- samples of minerals
- poster board
- scale
- measuring cup
- water

2. As a group, prepare a poster to describe your classification system. In the space below, sketch your ideas for the poster.

## Drawing Conclusions

1. What characteristics did you use as a basis for your classification system?
   _____
   _____

2. Did other groups use different characteristics? If so, describe them.
   _____
   _____

3. After reviewing the posters of other groups, would you like to change your group's system? Explain.
   _____
   _____

Name_____ Date_____

# Growing Crystals

**QUICK LAB**
FOR SCHOOL OR HOME
Lesson 3

**Hypothesize**  How can you watch crystals grow?
Write a **Hypothesis**:

_____
_____

## Procedure

Your teacher will put a cup of hot water onto a counter for you.

**BE CAREFUL!** Wear goggles. Use a kitchen mitt if you need to hold or move the cup. Don't touch the hot water.

1. Use a plastic spoon to gradually add small amounts of salt to the water. Stir. Keep adding and stirring until no more will dissolve.

2. Tie one end of a 15-cm piece of string to a crystal of rock salt. Tie the other end to a pencil. Lay the pencil across the cup so that the crystal hangs in the hot salt water without touching the sides or bottom.

3. **Observe** Observe the setup for several days. Record what you see.

_____
_____

### Materials

- foam cup half-filled with hot water
- granulated table salt
- 2 plastic spoons
- crystal of rock salt
- string (about 15 cm)
- pencil
- goggles

Unit C · Earth and Its Resources        Use with textbook page C37

Name _____ Date _____

**QUICK LAB** FOR SCHOOL OR HOME
Lesson 3

**Drawing Conclusions**

4. Did any crystals grow? If so, did they have many shapes or just one? Explain your answer. If not, how would you change what you did if you tried again?

_____

_____

_____

5. **Going Further** Grow needle-shaped crystals. Dissolve Epsom salt in water. Line a shallow dish with black construction paper so the crystals will be more visible. Place a very small amount of the solution in the dish. Set the dish where it will not be disturbed. Check the dish after one day for long needle-shaped crystals.

Name_____ Date_____

## Explore Activity
**Lesson 4**

# How Are Rocks Alike and Different?

**Hypothesize** Are rocks all alike? Are they different? If so, how? Write a **Hypothesis:**

_____
_____

### Materials
- samples of rocks
- clear tape
- red marker
- hand lens
- copper wire
- streak plate
- balance
- metric ruler
- calculator

## Procedure: Design Your Own

1. Use the tape to number each sample in a group of rocks.

2. **Classify** Find a way to sort the group into smaller groups. Determine which properties you will use. Group the rocks that share one or more properties. Your fingernail, the copper wire, and the edge of a streak plate are tools you might use. Scratch gently. Record your results.

_____
_____
_____
_____

3. **Use Numbers** You might estimate the density of each sample. Use a balance to find the mass. Use a metric ruler to estimate the length, width, and height.
**Length x width x height = volume; Density = mass ÷ volume**

_____
_____

## Drawing Conclusions

1. How were you able to make smaller groups? Give supporting details from the notes you recorded.

_____
_____

Unit C · Earth and Its Resources    Use with textbook page C41    95

Name_____ Date_____

**Explore Activity**
**Lesson 4**

2. Could you find more than one way to sort rocks into groups? Give examples of how rocks from two different smaller groups may have a property in common.

   _____
   _____

3. **Communicate** Share your results with others. Compare your systems for sorting the rocks.

   _____

4. **FURTHER INQUIRY** Infer How might a sample be useful based on the properties that you observed?

   _____
   _____

**Inquiry**

Think of your own questions that you might like to test. How can you learn more about your rock samples?

My Question Is:

_____
_____

How Can I Test It:

_____
_____

My Results Are

_____
_____

Unit C · Earth and Its Resources        Use with textbook page C41

Name_____ Date_____

**Alternative Explore**
Lesson 4

# Display Types of Rocks

## Procedure

1. Observe the rocks your teacher gave you. Use the hand lens to get a close look.

2. With your group, discuss ways that the rocks are alike and ways that they are different. Brainstorm ways to group the rocks by similarities. Record your group's ideas about grouping the rocks.

_____
_____
_____
_____

3. With your group, plan a display or poster to show how you grouped the rocks. Label each group to show the characteristics of the group.

**Materials**
- rock samples
- hand lens
- art supplies
- poster board

## Drawing Conclusions

1. What characteristics did you choose to use as a basis for grouping the rocks?

_____
_____

2. How did you show your decisions, with a poster or a display?

_____

3. Did other groups use different characteristics to group their rocks? Describe another group's system that you think was a good idea.

_____
_____
_____

Name_____ Date_____

**Inquiry Skill Builder**
Lesson 4

# Define Based on Observations

## Defining Soil

Earth's crust is made up of rocks and minerals. However, to get to the rocks, you usually have to dig through layers of soil.

Soil looks different at different places. It has different properties. Soil can be sandy. It can be moist.

Just what is soil? Make some observations. Write a definition that fits your observations.

### Materials

- moist soil sample in a plastic bag
- sand sample in a plastic bag
- hand lens
- 2 cups
- 2 plastic spoons

## Procedure

1. **Observe** Use a hand lens to examine a sample of moist soil. What materials can you find? How do their sizes compare? Write a description.

   _____
   _____

2. Some soils are more like sand. How does a sample of sand compare with your moist soil sample?

   _____

3. **Use Variables** Which sample absorbs water more quickly? Fill a cup halfway with sand and another with moist soil. Pour a spoonful of water in each at the same time.

   _____

4. **Experiment** Which absorbs more water? Make a prediction. Find a way to test your prediction.

   _____
   _____

Unit C · Earth and Its Resources — Use with textbook page C48

Name_____ Date_____

**Inquiry Skill Builder**
**Lesson 4**

5. **Experiment** Make any other observations. Look for other differences.
   _____
   _____

## Drawing Conclusions

1. Based on your observations, what is soil made up of?
   _____
   _____
   _____

2. How may soils differ?
   _____
   _____
   _____
   _____
   _____
   _____

3. **Define** Write a definition for *soil*. Take into account all your observations.
   _____
   _____
   _____

Name_____ Date_____

## Explore Activity
### Lesson 5

# What Makes Air Dirty?

**Hypothesize** What kinds of pollutants are in the air that can make it look as it does in the picture on the opening page of Lesson 5 of your textbook?

Write a **Hypothesis**:

_____
_____
_____
_____

**Materials**
- 12 cardboard strips
- petroleum jelly
- plastic knife
- transparent tape
- string
- hand lens
- metric ruler
- marker

### Procedure

1. Make square "frames" by taping together the corners of four cardboard strips. Make three frames, and label them A, B, and C. Tie a 30-cm string to a corner of each frame.

2. Stretch and attach three strips of tape across each frame, with all sticky sides facing the same way. Use a plastic knife to spread a thin coat of petroleum jelly across each sticky side.

3. **Use Variables** Hang the frames in different places to try to collect pollutants. Decide on places indoors or outdoors. Be sure to tell a parent or teacher where.

4. **Observe** Observe each frame over four days. On a separate page, record the weather and air condition each day.

5. **Measure** Collect the frames. Observe the sticky sides with a hand lens and metric ruler to compare particles. Record your observations.

_____
_____
_____
_____

Name_____ Date_____

**Explore Activity**
**Lesson 5**

## Drawing Conclusions

1. **Interpret Data** How did the frames change over time? How did the hand lens and ruler help you describe any pollution?

   _____
   _____
   _____
   _____

2. **Communicate** Present your data in a graph to show differences in amounts. Use a separate piece of paper.

3. **FURTHER INQUIRY** **Use Variables** What kinds of pollutants would your frames not collect? How might you design a collector for them? How might you extend this activity over different periods of time?

   _____
   _____
   _____

## Inquiry

Think of other questions that you might like to test. What type of particles do common air filters trap?

My Question Is:

_____
_____

How I Can Test It:

_____
_____

My Results Are:

_____

Unit C · Earth and Its Resources    Use with textbook page C61

Name_____ Date_____

**Alternative Explore**
Lesson 5

# Water Collector

## Procedure

1. Place a small amount of water in a container. Do not cover the container.
2. Place your container in a location where it won't be disturbed. Possible locations include a garden, the classroom, an office, or a room at home.
3. Leave the container in place until the water evaporates.
4. After the water has evaporated, examine the bottom of the container. What do you see?

   _____
   _____

**Materials**
- gallon containers
- water

## Drawing Conclusions

1. Where did you place your container?

   _____

2. How long did it take for the water to evaporate?

   _____

3. Observe your classmates' containers. Did all the containers contain the same type of material? Explain any differences you saw.

   _____
   _____
   _____
   _____

4. Where did the material that you found in the container come from?

   _____
   _____

Name_____ Date_____

# Acids

**For School or Home**
**Lesson 5**

**Hypothesize** How can acid rain change a rock?
Write a **Hypothesis**:
_____
_____

Your teacher will give you a stick of chalk and some rock samples.

## Procedure

**BE CAREFUL!** Wear goggles.

1. **Use Variables** Break a stick of chalk into smaller pieces. Place some small pieces in a plastic cup. Place each rock sample in its own cup. Slowly pour vinegar into each cup to cover each object.

2. Cover each cup using plastic wrap and a rubber band to help keep the vinegar from evaporating.

3. **Observe** Watch for any changes in the chalk and the rocks. Watch for several minutes and then at later times in the day. Record your observations.

_____
_____
_____
_____
_____

### Materials

- chalk
- limestone and other rock samples
- vinegar (a mild acid)
- plastic cups
- goggles
- plastic wrap
- rubber bands
- plastic knife

Unit C · Earth and Its Resources          Use with textbook page C65

Name _____ Date _____

**Quick Lab — For School or Home — Lesson 5**

## Drawing Conclusions

4. Vinegar is a mild acid. How did it change the chalk?

   _____
   _____
   _____

5. Do all rocks change the same way? Explain based on your results.

   _____
   _____

6. **Going Further** Acid rain causes metals to deteriorate more rapidly. This can be simulated using steel wool and vinegar. Write and conduct an experiment to simulate more rapid deterioration of metal by acid rain.

   My Hypothesis Is:

   _____
   _____

   My Experiment Is:

   _____
   _____
   _____
   _____
   _____

   My Results Are:

   _____
   _____

Unit C · Earth and Its Resources          Use with textbook page C65

Name_____ Date_____

# Investigate How to Make Salt Water Usable

**Hypothesize** How can water with something dissolved in it be changed into fresh water? Test your ideas.

Write a **Hypothesis**:

_____
_____

**Materials**
- tea bag
- deep pan
- plastic cup
- saucer (or petri dish)
- large, clear bowl or container
- water

## Procedure

1. **Make a Model** Keep a tea bag in a cup of water until the water is orange.

2. **Make a Model** Place a pan where there is strong light (sunlight, if possible). Pour some tea water into the saucer. Put the saucer in the pan. Cover the saucer with a large bowl.

3. **Observe** Look at the bowl and pan several times during the day and the next day. Note any water you see on the bowl or in the pan. Record your observations.

_____
_____
_____
_____
_____

## Drawing Conclusions

1. How was the water that collected in the bowl and in the pan different from the tea water?

_____
_____

Unit C · Earth and Its Resources        Use with textbook page C71

Name_____ Date_____

**Explore Activity Lesson 6**

2. **Infer** What do you think caused the water to collect in the bowl and pan?

   _____

3. How does this model represent what might happen to salt water, the water of Earth's oceans?

   _____
   _____
   _____

4. **Use Variables** How long did it take for water to collect in the bowl and pan? How might this process be speeded up?

   _____
   _____

5. **FURTHER INQUIRY** **Communicate** Suppose you added salt to the water instead of using a tea bag. How could you tell if the salt had been removed? Try it. Might this model work as a way to get fresh water from ocean water?

   _____
   _____

### Inquiry

Think of your own questions that you might like to test. What other substances can your model remove from water?

My Question Is:

_____

How I Can Test It:

_____
_____

My Results Are:

_____
_____

106  Unit C · Earth and Its Resources    Use with textbook page C71

Name _____ Date _____

**Alternative Explore**
Lesson 6

# What's Left?

## Procedure

**BE CAREFUL!** Wear goggles.

1. Pour some saltwater solution into the pie pan.
2. Place the pie pan near a heat source, such as sunlight or a strong lamp.
3. Check the dish every hour to see if the water has evaporated.
4. When the water has evaporated, observe the material in the dish. Use a hand lens to get a closer look. Record your observations. Draw what you see through the magnifying lens in the space below.

_____
_____

**Materials**
- saltwater solution
- hand lens
- heat source
- goggles
- pie pan

## Drawing Conclusions

1. What would be left if all the water in the oceans evaporated?

_____
_____

2. Why would it be useful to collect the water as it evaporates?

_____
_____
_____

Unit C · Earth and Its Resources       Use with TE textbook page C71

Name _____  Date _____

**Inquiry Skill Builder**
**Lesson 6**

# Form a Hypothesis

## How Do Wastes from Land Get into Lakes and Rivers?

In seeking an answer to a question, the first thing you might do is find out as much as possible. You make observations. You might look up information.

Next, you would think of an explanation for these observations. That explanation is a hypothesis. It may be stated as an "If . . . then" sentence. "If water runs over the land where garbage is dumped, then . . ." Sometimes you can test a hypothesis by making and observing a model.

### Materials
- soil
- food color
- foam bits
- 2 aluminum pans
- water
- 2 textbooks

## Procedure

1. **Hypothesize** Write a hypothesis to answer the question above.

   _____
   _____
   _____

2. **Make a Model** Pack moist soil to fill one-half (one side) of one aluminum pan. As you pack the soil, add 10–20 drops of food color to the soil just below the surface. Sprinkle crumbled bits of foam over the top.

3. **Experiment** Use two books to tilt the pan with the soil side up. Place the lower edge of the soil-filled pan in the other pan. Pour water over the uppermost edge of the pan. Describe what happens. Let your model stand for some time, and observe it again.

   _____
   _____
   _____
   _____

Unit C · Earth and Its Resources          Use with textbook page C77

Name_____ Date_____

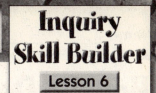

## Drawing Conclusions

1. How does this model represent wastes on land?

   _____
   _____
   _____
   _____

2. Based on the model, how do wastes from land get into water? Does the model support your hypothesis? Explain.

   _____
   _____
   _____
   _____

3. **Hypothesize** How can some wastes be removed from water? Form a hypothesis, and test your ideas.

   My Hypothesis Is:

   _____
   _____

   My Procedure Is:

   _____
   _____

   My Results Are:

   _____
   _____

Name _____ Date _____

# Explore Activity
## Lesson 7

# How Do Ocean and Fresh Water Compare?

**Hypothesize** How does the density of fresh water compare to the density of salt water? Test your ideas.

Write a **Hypothesis**:
_____
_____

### Materials
- 3 small plastic cups
- "ocean water"
- "fresh water"
- clear-plastic straw
- waterproof marking pen
- wax paper
- ruler

## Procedure

1. Spread wax paper on your desk before you begin to work.

2. **Predict** What happens when you mix fresh and ocean water?

_____
_____

3. **Experiment** From the bottom of the straw, make a mark every 4 cm. Gently place the bottom of the straw 4 cm under the surface of the "ocean water." Seal the top of the straw with your finger. With your finger still sealing the straw, lift it out of the water. Keeping your finger on top of the straw, place the bottom of the straw 8 cm down in the "fresh water." Lift your finger off the straw, then put it back again and lift the straw out of the water.

4. **Observe** What happened? Record the results. Now try it again, starting with "fresh water" first. Observe and record what happens.

_____
_____
_____
_____

Unit C · Earth and Its Resources    Use with textbook page C83

Name_____ Date_____

## Drawing Conclusions

1. **Communicate** Which liquid combinations mixed in the straw and which made layers?
   _____
   _____

2. **FURTHER INQUIRY** **Experiment** Make a third liquid by mixing equal parts of ocean water and fresh water. How will the mixture compare to fresh and ocean water? Make your prediction, then test it.
   _____
   _____

**Inquiry**

Think of your own question that you might like to test. Will the layers created by adding fresh water on top of salt water stay separate indefinitely?

My Question Is:
_____
_____

How I Can Test It:
_____
_____

My Results Are:
_____
_____

Unit C · Earth and Its Resources     Use with textbook page C83

Name_____ Date_____

# Salt Water Density

## Procedure

1. Pour the cupful of colored salt water into the fresh water along the side of the bowl.

2. Observe and record what happens.

   _____

   _____

3. Now reverse the experiment. Pour the cupful of colored fresh water into the salt water along the side of the bowl.

4. Observe and record what happens.

   _____

**Materials**
- water
- salt
- food coloring
- cup
- deep glass bowl
- teaspoon

## Drawing Conclusions

1. Which is denser, fresh water or salt water? Why?

   _____

   _____

2. What do you think would happen if you mixed salt into the fresh water and repeated the experiment?

   _____

   _____

3. How could the salt water be made even denser?

   _____

Name_____ Date_____

# Salt Water and Fresh Water

**Hypothesize** How will salt water and fresh water differ in the way that they affect a floating object?
Write a **Hypothesis**:

_____

_____

**Materials**
- jar
- pencil with eraser
- thumbtack
- "fresh water"
- "salt water"
- waterproof marker
- ruler

## Procedure

1. Fill a jar with fresh water to about 1 cm from the top. Carefully push a thumbtack into the center of the eraser of a pencil.

2. **Observe** Place the pencil, eraser side down, in the water. Let go. What happens?

_____

3. **Measure** Using a waterproof marker, mark the pencil to show where the water line is. Use a ruler to measure the length of the pencil above the water mark. Record this measurement.

_____

_____

4. Fill the jar with salt water. Repeat steps 2–3. Record your results. Compare with your results for fresh water.

_____

Unit C · Earth and Its Resources   Use with textbook page C85

Name _____ Date _____

**QUICK LAB FOR SCHOOL OR HOME**
Lesson 7

## Drawing Conclusions

5. **Predict** What do you think will happen if you add a tablespoon of salt to your salt water? Test your prediction.

   _____
   _____
   _____
   _____

6. **Going Further** Would it be easier to swim in a freshwater lake or in a saltwater lake? Explain your answer.

   _____
   _____
   _____
   _____

Name_____ Date_____

# How Do People Use Energy?

**Hypothesize** How many different ways do you use energy each day? How can you use less energy? Test your ideas.

Write a **Hypothesis**:

_____

_____

## Procedure

1. **Communicate** Make a list of all the different ways you use energy.

   _____

   _____

2. Make a table listing all the kinds of energy you use in a day, how you use that energy, and how many hours you use each kind. Put your table on a separate sheet of paper.

## Drawing Conclusions

1. How many different ways do you use electricity each day? How many hours a day do you use electricity? What other sources of energy do you use? How many hours a day do you use each?

   _____

   _____

   _____

   _____

Name _____  Date _____

**Explore Activity**
**Lesson 8**

2. **Infer** Make a log to keep track of your energy use at home and at school. How can you use that information to help you make a plan to save energy?

_____
_____

3. **Use Numbers** If it costs you an average of ten cents an hour for the energy you use, how much would the energy you use cost each week? About how much would it cost each month?

_____
_____

4. **FURTHER INQUIRY** **Hypothesize** How can you use less electricity? How much money do you think you could save on energy use in a month? Design and carry out a test of your hypothesis.

_____
_____
_____
_____

**Inquiry**

Think of your own questions that you might like to test. Can you conserve energy by using less hot water?

My Question Is:
_____

How I Can Test It:
_____
_____

My Results Are:
_____
_____

Name_____ Date_____

**Alternative Explore**
Lesson 8

# Pool the Data

**Procedure**

1. List the ways members of your group use energy, such as watching TV, burning gasoline in a car you ride in, burning a fuel to heat your home. Write each way you use energy on a separate card.

2. After you have listed all the uses, ask each member how much time each day he or she uses energy in that way.

3. If times of use vary a large amount from one group member to another, estimate an average time. Add the time to each card.

4. Arrange the cards in order of time used, from most to least.

5. Prepare a poster listing the ways your group uses energy and the times that each use lasts. List the uses in order of time used.

**Materials**
- file cards
- chart paper
- marking pens

**Drawing Conclusions**

1. What way does your group use energy for the longest time?

   _____

2. Did your group list heating the home as a 24-hour-per-day activity? Not all systems use energy constantly. Ask a family member to help you figure out how much of the time the system in your home actively uses energy to produce heat. How does this information change your results?

   _____

   _____

3. Review the ways you use energy. Which ones would you be willing to cut down on in an effort to save energy?

   _____

Name_____ Date_____

# Fuel Supply

**Hypothesize** We are using fossil fuels at the rates shown in the table. How long will Earth's fossil fuel supply last? Write a **Hypothesis:**

_____
_____

## Procedure

This table shows how fast we are using up oil and natural gas.

| WORLD SUPPLY OF OIL AND NATURAL GAS (as of January 1, 1996) | |
|---|---|
| Oil | 1,007 billion barrels (1,007,000,000,000) |
| Natural gas | 4,900 trillion cubic feet |
| **WORLD USE OF OIL AND NATURAL GAS FOR 1995** | |
| Oil | about 70 million barrels a day (70,000,000) |
| Natural gas | about 78 trillion cubic feet |

1. **Observe** Examine the data in the table.
2. **Communicate** Draw a graph in the space below showing how long the fossil fuels we know about will last, based on the data in the table.

Unit C · Earth and Its Resources     Use with textbook page C103

Name_____ Date_____

## Drawing Conclusions

3. **Infer** How long will it be until we run out of each type of fossil fuel? Assume that the rate of use remains the same.

   _____

4. **Going Further** Sources of energy other than fossil fuels are becoming more common. Write and conduct an experiment to learn about usage of alternative energy sources.

   My Hypothesis Is:

   _____

   _____

   My Experiment Is:

   _____

   _____

   My Results Are:

   _____

   _____

Unit C · Earth and Its Resources          Use with textbook page C103

Name_____ Date_____

**Explore Activity**
**Lesson 1**

# How Are Earth and the Sun Held Together?

**Hypothesize** How does a force hold Earth around the Sun? What would happen if the force let go? Write a **Hypothesis:**

_____
_____

**Materials**
- clay
- string
- scissors
- meterstick
- goggles

## Procedure

**BE CAREFUL!** Wear goggles. Twirl the model close to the ground.

1. **Make a Model** Cut a 40-cm length of string. Wrap it around a small, round lump of clay in several directions. Tie the ends to make a tight knot. Measure 60 cm of string, and tie it to the string around the ball.

2. **Observe** Spin the ball of clay slowly—just fast enough to keep the string tight and keep the ball off the ground. Keep the ball close to the ground. Describe the path of the ball

   _____

3. **Experiment** At one point while spinning, let the string go. What happens? Describe the path of the ball of clay. Repeat until you get a clear picture of what happens.

   _____

## Drawing Conclusions

1. How did your model represent Earth and the Sun? What represented Earth? Where was the Sun located? How did you represent the force between them?

   _____
   _____
   _____

Name_____  Date_____

**Explore Activity Lesson 1**

2. **Infer** Explain what happened when you let the string go. Why do you think this happened?

   _____
   _____
   _____

3. **FURTHER INQUIRY** **Use Variables** How would your results change if the mass of the clay was doubled? Tripled? How does the mass affect the pull on the string? Make a prediction. Try it.

   _____
   _____
   _____

### Inquiry

Think of your own questions that you might like to test. What other conditions affect the pull on the string or the path of the released ball of clay?

My Question Is:

_____
_____

How I Can Test It:

_____
_____

My Results Are:

_____
_____

Unit D · Astronomy, Weather and Climate          Use with textbook page D5

Name_____ Date_____

**Alternative Explore**
**Lesson 1**

# Motion of a Planet

## Procedure

**Materials**
- paper plate
- scissors
- marble

1. Place a paper plate on a table. Place a marble in the rut near the rim of the plate.

2. Lift one edge of the plate slightly. Observe what happens. Record your observations.

   _____
   _____

3. Cut the paper plate in half. Place one half of the plate on the table.

4. Place a marble in the rut near the rim of the plate. Lift the edge of the plate slightly. Observe what happens. Record your observations. Draw the path of the marble in the space below. Show the paper plate in your drawing.

   _____
   _____

## Drawing Conclusions

1. How is the motion of the marble in the whole paper plate similar to Earth's motion around the Sun?

   _____

2. How does cutting the plate in half change the forces on the marble?

   _____
   _____

3. What would have to happen for Earth's orbit to be like the path of the marble in step 4?

   _____

Unit D · Astronomy, Weather and Climate      Use with textbook page D5

Name_____ Date_____

# Orbit Times

**Hypothesize** What does the length of time for an orbit depend on? Test your ideas. Write a **Hypothesis:**

_____
_____

**Materials**
- several sheets of graph paper

## Procedure

1. **Communicate** Use graph paper. Draw a bar graph to compare the revolution times for the planets. The vertical axis of the graph represents time. Decide how much time each square on the paper represents. The horizontal axis represents the planets. How many pieces of graph paper will you need? Write your description.

   _____

## Drawing Conclusions

2. **Interpret Data** Based on your graph and the data table, what relationship can you find between the length of the year (time) and the planet's location in the solar system?

   _____
   _____

3. How could you change your graph to show the relationship even better? What might your new graph reveal?

   _____
   _____
   _____
   _____

4. **Use Variables** Draw a line graph on a separate sheet of paper. Let the vertical axis represent time for a complete orbit. Let the horizontal axis represent distance to the Sun. Label each point with the name of its planet.

Unit D · Astronomy, Weather and Climate        Use with textbook page D7

Name _____ Date _____

**QUICK LAB**
FOR SCHOOL OR HOME
Lesson 1

5. **Going Further** Do all the planets travel at the same speed? Calculate each planet's speed. First calculate the distance traveled by treating the orbits as circular and calculating the circumference of the circle ($2\pi r$ where $r$ is the distance to the Sun). Convert the orbit time from days to hours by multiplying the number of days by 24 hours. Calculate the speed by dividing the distance traveled by the hours it takes for a complete orbit around the Sun. Multiply the speed in million km/hour by 1000 to report it in thousand km/hour.

| PLANET | Distance Traveled (million km) | Orbit Time (hours) | Speed (thousand km/hour) |
|---|---|---|---|
| Mercury | | | |
| Venus | | | |
| Earth | | | |
| Mars | | | |
| Jupiter | | | |
| Saturn | | | |
| Uranus | | | |
| Neptune | | | |

On a piece of graph paper, plot the average distance to the Sun versus the planet's orbital speed. What is the relationship between a planet's orbital speed and its distance to the Sun? Explain this relationship.

_____

Name_____ Date_____

# How Do the Distances Between Planets Compare?

**Hypothesize** Can you tell the distance between you and other planets? Write a **Hypothesis**:

_____
_____

**Materials**
- roll of paper towels
- markers
- tape (optional)
- ruler

## Procedure

1. **Use Numbers** Study the chart "Planet Distances From the Sun" on page D15. Each distance is expressed in Astronomical Units (A.U.). One A.U. is the distance from Earth to the Sun. How far from the Sun is Mars? Pluto?

_____

2. **Make a Model** Let the width of one paper towel be one A.U. Lay out the length of paper towels you need to show the distance from the Sun to Pluto. Measure and mark the location of each planet.

_____

## Drawing Conclusions

1. **Interpret Data** Describe the general spacing of the planets starting from the Sun.

_____
_____
_____
_____

Unit D · Astronomy, Weather and Climate    Use with textbook page D15

Name _____ Date _____

**Explore Activity**
**Lesson 2**

2. **Use Numbers** Earth is about 150 million km (93 million miles) from the Sun. How far away from the Sun is Mercury? Pluto?

   _____
   _____
   _____

3. **Use Numbers** It takes 8 minutes for light to travel from the Sun to Earth. How long does it take for light to travel to Jupiter? To Pluto?

   _____
   _____

4. **FURTHER INQUIRY** **Make a Model** In your model, all the planets are in a straight line. Actually, they are scattered in different places in their orbits. How can you change your model to be more accurate? Make a plan and try it.

   _____
   _____
   _____

**Inquiry**

Think of your own question related to the distances between planets.

My Question Is:

_____
_____

How Can I Test It:

_____
_____

My Results Are:

_____
_____

Name _____ Date _____

# Planet Distance Model

## Procedure

1. Take nine sticky-notes and write each name of the planet on them and write one for the Sun. Place a toothpick in each of the Styrofoam balls. Place the sticky-notes in the order of the planets on the toothpicks.

2. Take the yardstick and place each planet in a straight line according to the Astronomical Unit (A.U.) chart on page D15. Measure them on the floor where 1 yard = 1 A.U. Use the Sun as your starting point.

**Materials**
- several sizes of Styrofoam balls
- yardstick
- toothpicks
- sticky-notes

## Drawing Conclusions

1. Some planets are closer to the Sun than others. Making this model gives you an idea of how far apart they are from each other. Which planet takes the least amount of time to rotate the Earth? The least?

   _____
   _____

2. In the solar system, the planets are scattered along their orbit. Draw below what the planets would actually look like in their orbit.

Name _____ Date _____

**Inquiry Skill Builder**
Lesson 2

# Make a Model

## Making a Model of the Solar System

In this activity you will make a model to compare the sizes of the planets in the solar system. The table "Comparing a Planet's Radius with Earth's," will tell you how the radius of each "model planet" you make would compare to Earth's radius.

### Materials
- construction paper
- white paper
- pencil
- string 25 cm long
- metric ruler
- colored markers or colored pencils
- tape

## Procedure

1. **Use Numbers** Look at the table "Comparing a Planet's Radius with Earth's" on page D17. How much bigger is Jupiter's radius than Earth's radius? How much smaller is Mars' radius than Earth's?

   _____

   _____

2. **Measure** Let your model Earth's radius be 1 cm. Using this scale, how big would you need to make the radius of Jupiter? How big would you need to make the radius of Mars?

   _____

3. **Make a Model** Draw a model Earth with a 1 cm radius. Cut out your model. Repeat this process for each planet.

Unit D · Astronomy, Weather and Climate    Use with textbook page D17

**Drawing Conclusions**

1. **Compare** Look at the sizes of your model planets. Which planets are almost the same size?

   _____

   _____

2. **Compare** The radius given for Saturn was given for the planet only, without its rings. If Saturn's radius equaled the distance from its center to its outermost ring, Saturn's radius would be over 28.5 times larger than Earth's radius. How much larger then would Saturn's radius be compared to Jupiter's?

   _____

Name_____ Date_____

**Explore Activity** — Lesson 3

# Does the Sun's Angle Matter?

**Hypothesize** How does the angle at which the Sun's energy hits Earth affect the warming of Earth? Write a **Hypothesis:**

_____

_____

## Procedure

**BE CAREFUL!** Do not look into the lamplight. Prop up a foam bowl, using a plate or clay, to shield your eyes from the light.

1. Place a thermometer onto each of the three blocks, as shown. Cover each with black paper. Put blocks 20 cm from the light bulb, level with its filament (curly wire).

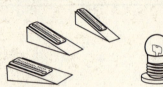

2. **Observe** Measure the starting temperature at each block. Record the temperatures.

_____

3. **Predict** What will happen when the lamp is turned on? Turn the lamp on. Record the temperature at each block every two minutes for ten minutes, in a data table on another sheet of paper.

_____

_____

4. **Communicate** On another sheet of paper, make a line graph showing the change in temperature at each block over time.

5. **Use Variables** Repeat the activity with white paper.

### Materials

- 3 thermometers
- centimeter ruler
- stopwatch
- triangular blocks
- scissors
- foam bowl
- black paper
- tape
- clay
- white paper
- 150-W clear-bulb lamp

Name _____ Date _____

**Explore Activity — Lesson 3**

## Drawing Conclusions

1. **Communicate** Which block's surface was warmed most by the lamplight? Which block's surface was warmed the least?

   _____
   _____

2. **Infer** How does the angle at which light hits a surface affect how much the surface is heated? How does the surface color affect how much it is heated?

   _____
   _____
   _____

3. **FURTHER INQUIRY** **Experiment** What other factors might affect how much a surface is warmed by sunlight? How would you test your ideas?

   _____
   _____
   _____

**Inquiry**

Think of your own questions that you might like to test. What other factors might affect temperature on different parts of Earth?

My Question Is:

_____

How I Can Test It:

_____

My Results Are:

_____

Unit D · Astronomy, Weather and Climate  Use with textbook page D29

Name_____ Date_____

**Alternative Explore**
**Lesson 3**

# Cold at the Poles

## Procedure

1. Darken the classroom by shutting off the lights and closing the blinds.
2. Place a globe on your desk.
3. Holding a flashlight, stand one or two paces from the globe.
4. Turn on the flashlight and shine it onto the globe.
5. Hold the flashlight level with the equator. Observe how the light strikes the equator and record your observations.

_____
_____

6. **Observe** How does the light strike the north and south poles? Record your observations.

_____
_____

**Materials**
- globe
- flashlight

## Drawing Conclusions

1. What do the globe and flashlight represent?

_____
_____

2. **Compare and Contrast** How does sunlight strike Earth at the equator and at the poles?

_____
_____

3. Why do you think temperatures near the equator are warmer than those near the poles?

_____
_____

Name_____ Date_____

# Investigating Angles

**Hypothesize** Why does the angle of insolation cause a difference in warming? Write a **Hypothesis**:

_____
_____
_____
_____
_____

**Materials**
- flashlight
- modeling clay
- 3 toothpicks
- sheet of graph paper
- ruler

## Procedure

1. Fold a sheet of graph paper lengthwise in three equal parts. Put a small lump of clay in the middle of each part. Stand a toothpick straight up in each lump of clay.

2. Hold a flashlight directly over the first toothpick. Have a partner trace a line around the circle of light and trace the toothpick shadow.

3. **Use Variables** Repeat step 2 for the other two toothpicks, changing only the angle of the flashlight.

4. **Measure** Count the number of boxes in each circle. Measure the lengths of the toothpick shadows. Record your results in the table below.

| Toothpick | Number of Boxes | Length of Shadow |
|---|---|---|
| 1 | | |
| 2 | | |
| 3 | | |

Unit D · Astronomy, Weather and Climate · Use with textbook page D31

Name_____ Date_____

**Quick Lab — For School or Home — Lesson 3**

## Drawing Conclusions

5. **Infer** How is the length of the shadows related to the angle?

   _____
   _____

6. **Infer** How is the number of boxes in the circle related to the angle?

   _____
   _____

7. **Going Further** How does the angle of insolation affect where you live? Compare and contrast the local climate with the climate at the equator and at Earth's poles. Write and conduct an experiment.

   My Hypothesis Is:

   _____
   _____
   _____

   My Experiment Is:

   _____
   _____
   _____
   _____
   _____

   My Results Are:

   _____
   _____
   _____
   _____

Name_____ Date_____

# Where Does the Puddle Come From?

**Materials**
- plastic cups
- ice
- paper towels
- food coloring
- thermometer
- goggles

**Hypothesize** Think about putting a cold glass of lemonade on a table on a hot, humid day. Moisture forms on the side of the glass and in a puddle around the bottom. Where does the moisture come from? How might you design an experiment to test your ideas? Write a **Hypothesis**:

_____
_____

## Procedure: Design Your Own

**BE CAREFUL!** Wear goggles.

1. **Form a Hypothesis** At the top of this page, write down your idea about why a puddle forms around a frosty drink. Where do you think the puddle came from?

2. **Experiment** Describe what you would do to test your idea. How would your test support or reject your idea?

_____
_____
_____
_____

3. **Communicate** On another sheet of paper draw a diagram showing how you would use the materials. Keep a record of your observations.

## Drawing Conclusions

1. **Communicate** Describe the results of your investigation.

_____
_____

2. **Communicate** What evidence did you gather? Explain what happened.

_____
_____

Unit D · Astronomy, Weather and Climate     Use with textbook page D37

Name_____  Date_____

**Explore Activity**
**Lesson 4**

3. **Infer** How does this evidence support or reject your explanation?

_____
_____

4. **FURTHER INQUIRY** Use Variables Do you get the same results on a cool day as on a warm day? Do you get the same results on a humid day as on a dry day? Investigate to test your hypothesis.

_____
_____
_____
_____
_____

**Inquiry**

Think of your own questions that you might like to test. How might temperature affect the results of this experiment?

My Question Is:

_____
_____

How I Can Test It:

_____
_____

My Results Are:

_____
_____

Name_____ Date_____

# Where'd It Go?

## Procedure

1. Pour an equal amount of water into each of the jars.
2. Tightly screw the lid onto one of the jars.
3. With a wax pencil, mark the level of water in each jar.
4. Leave the jars where they will be undisturbed overnight.
5. The next day, observe the water level in each jar and record your observations.

   _____
   _____

6. Record any other observations you made.

   _____
   _____

**Materials**
- two jars, one with a lid
- wax pencil
- water

## Drawing Conclusions

1. What do you think happened to the water in the open jar?

   _____
   _____
   _____

2. Did you observe the same results in both jars? Explain why or why not.

   _____
   _____
   _____
   _____
   _____

Unit D · Astronomy, Weather and Climate        Use with textbook page D37

Name_____ Date_____

**QUICK LAB**
FOR SCHOOL OR HOME
Lesson 4

# Transpiration

**Hypothesize** What evidence can you find for transpiration? Write a **Hypothesis:**

_____

_____

**Materials**
- potted houseplant (geraniums work well)
- transparent plastic bag

## Procedure

1. Place a clear-plastic bag completely over a houseplant. Tie the bag tightly around the base of the stem. Do not put the soil-filled pot into the bag.

2. **Observe** Place the plant in a sunny location. Observe it several times a day. When you are done, remove the plastic bag from the plant.

_____

_____

## Drawing Conclusions

3. **Communicate** Describe what you see on the inside of the bag. How can you explain what happened?

_____

_____

_____

4. **Draw Conclusions** *Transpiration* sounds like *perspiration*—sweating. How might the two processes be alike?

_____

_____

5. **Predict** How would your results vary if you put the plant in the shade?

_____

_____

Unit D · Astronomy, Weather and Climate

Name_____ Date_____

6. **Going Further** How do you think temperature might affect transpiration? How would you set up a test?

My Hypothesis Is:

_____
_____

My Experiment Is:

_____
_____

My Results Are:

_____
_____

Name_____ Date_____

**Explore Activity**
Lesson 5

# How Do Clouds Form?

**Hypothesize** Sometimes the sky is full of clouds. Sometimes there are no clouds at all. Why? What makes a cloud form? What do evaporation and condensation have to do with it? Write a **Hypothesis**:

_____
_____
_____

**Materials**
- hot tap water
- 2 identical clear containers
- mug
- 3 ice cubes

Watch what can happen when you cool off some air.

## Procedure

**BE CAREFUL!** Be careful handling the hot water.

1. Chill container 1 by putting it in a refrigerator or on ice for about ten minutes.
2. Fill a mug with hot water.
3. **Make a Model** Fill container 2 with hot water. Place empty cold container 1 upside down on top of container 2 with the water. Fit the mouths together carefully. Place the ice cubes on top of container 1.
4. **Observe** Record your observations on another sheet of paper.

## Drawing Conclusions

1. **Communicate** What did you observe?

_____
_____

2. **Communicate** Where did this take place?

_____
_____

Name_____ Date_____

3. **Infer** Where did the water come from? Explain what made it happen.

_____
_____
_____
_____
_____
_____

4. **FURTHER INQUIRY** **Infer** Do clouds form better in dry or moist air? Conduct an experiment to test your inference. What materials will you need? What will you do?

_____
_____
_____
_____
_____
_____
_____
_____

### Inquiry

Think of your own questions that you might like to test. Do dry conditions affect clouds?

My Question Is:

_____
_____

How I Can Test It:

_____
_____
_____

My Results Are:

_____

Name_____  Date_____

**Alternative Explore**
Lesson 5

# Make a Cloud

## Procedure

**BE CAREFUL!** Keep away from the tea kettle spout and steam to avoid getting burned. Wait until the tea kettle cools before handling it.

1. Fill the tea kettle about half way with water.
2. Plug in the tea kettle, turn it on (if necessary), and wait for the water to start boiling.
3. **Observe** Observe the spout of the tea kettle and record your observations.

   _____
   _____

4. Use the mitt and tongs to hold the metal pie plate above the spout.
5. **Observe** Observe what happens and record your observations.

   _____
   _____

6. Turn off and unplug the tea kettle

### Materials
- electric tea kettle
- water
- metal pie plate
- tongs
- oven mitt

## Drawing Conclusions

1. What happened when the water in the kettle boiled? Explain your observations.

   _____
   _____
   _____

2. When you held up the pie pan, where did a "cloud" form? Why did it form there?

   _____
   _____
   _____

Name_____ Date_____

# Feel the Humidity

**Quick Lab — FOR SCHOOL OR HOME — Lesson 5**

**Hypothesize** Why do you feel warmer on a high humidity day? Write a **Hypothesis**:

_____
_____

## Procedure

**BE CAREFUL!** Be careful handling warm water.

1. **Observe** Use a thermometer to determine the air temperature.

   _____

2. Put the thermometer in cold water. Slowly add warm water until the water temperature matches the air temperature.

3. Wrap a 5-cm-square piece of old cotton cloth around the bulb of the thermometer. Gently hold it with a rubber band. Dampen the cloth in the water.

4. **Observe** Gently wave the thermometer in the air. Record temperatures every 30 seconds for 3 minutes in the table below.

### Materials
- 5-cm-square piece of old cotton cloth
- rubber band
- thermometer
- 1 c of warm water
- $\frac{1}{2}$ c of cold water

| Time | Temperature |
|---|---|
| 30 seconds | |
| 1 minute | |
| 1 minute, 30 seconds | |
| 2 minutes | |
| 2 minutes, 30 seconds | |
| 3 minutes | |

Unit D · Astronomy, Weather and Climate — Use with textbook page D48

Name_____ Date_____

**QUICK LAB**
FOR SCHOOL OR HOME
Lesson 5

**Drawing Conclusions**

5. **Infer** What happened to the temperature of the wet cloth? How does the cloth feel? Explain.

   _____
   _____

6. **Infer** If you try this experiment on a day that is humid and on a day that is dry will you get the same results? Explain.

   _____
   _____
   _____
   _____

7. **Going Further** In a school track meet, would sweating cool the runners more on a humid day or on a dry day? Write and conduct an experiment.

   My Hypothesis Is:

   _____
   _____

   My Experiment Is:

   _____
   _____

   My Results Are:

   _____
   _____

Name _____ Date _____

# What Can Change Air Pressure?

**Hypothesize** Air moves from one place to another because of differences in air pressure. What causes these differences? Make a model to test your ideas. Write a **Hypothesis:**

_____

_____

### Materials

- plastic jar with hole in bottom
- plastic sandwich bag
- rubber band
- masking tape

## Procedure

1. **Make a Model** Set up a jar-and-bag system as shown. Make sure the masking tape covers the hole in the jar. Have a partner place both hands on the jar and hold it firmly. Reach in and slowly pull up on the bottom of the bag. Describe what happens.

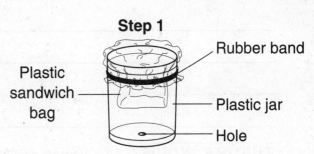

_____

2. **Experiment** Pull the small piece of tape off the hole in the bottom of the jar. Repeat step 1. Push in on the bag. Record your results.

_____

3. **Observe** Place some small bits of paper on the table. Hold the jar close to the table. Point the hole toward the bits of paper. Pull up on the bag, and observe and record what happens.

_____

4. **Experiment** Do just the opposite. Push the bag back into the jar. What happened?

_____

Unit D · Astronomy, Weather and Climate    Use with textbook page D53

Name_____ Date_____

## Drawing Conclusions

1. **Observe** What differences did you observe with the hole taped and with the tape removed?

   _____
   _____
   _____

2. **Infer** Explain what happened each time you pushed the bag back into the jar. How does this model show air pressure changes?

   _____
   _____
   _____
   _____
   _____

3. **FURTHER INQUIRY** **Use Variables** What happens to the amount of space air takes up if it is warmed? Use the model to test your hypothesis.

   _____

### Inquiry

Think of your own questions that you might like to test. Does air temperature affect air pressure?

My Question Is:

_____

How I Can Test It:

_____
_____

My Results Are:

_____

Name_____ Date_____

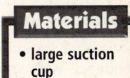

# Pushing Air

**Procedure**

1. Place the open end of the large suction cup against the chalkboard.

2. Push the stick attached to the cup as far as you can toward the chalkboard. Observe what happens and record your observations.

   _____
   _____

3. Let go of the stick. Observe what happens and record your observations.

   _____
   _____

**Materials**
- large suction cup

**Drawing Conclusions**

1. What happened to the air inside the suction cup when you pushed the stick in?

   _____
   _____

2. How did the pressure of the air inside the suction cup change after you pushed in and released the stick? Explain your answer.

   _____
   _____
   _____

3. Explain what happened to the suction cup after you let go of the stick.

   _____
   _____
   _____

Name _____     Date _____

**Inquiry Skill Builder**
**Lesson 6**

# Interpret Data

## A Weather Station Model

A weather station model includes temperature, cloud cover, air pressure, pressure tendency, wind speed, and wind direction. The circle is at the location of the station. You will interpret the data—use the information—from the weather use station models to answer questions and solve problems.

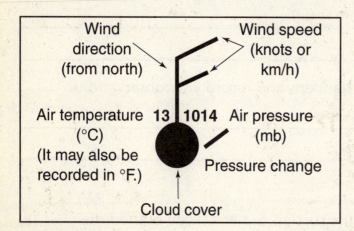

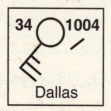
Dallas

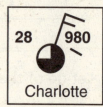
Charlotte

Oakland

Tampa

## Procedure

1. **Use Numbers** Look carefully at the Dallas weather station model. How fast is the wind blowing? What is the wind direction? Record your answers.

   _____

2. **Interpret Data** What other information does this weather station model give you?

   _____

   _____

3. Look at the other weather station models. On the next page, make a table recording weather conditions for each city.

Name _____ Date _____

# Inquiry Skill Builder
## Lesson 6

## Drawing Conlusions

1. Compare the information in the table you made with these station models. Which way is the information easier to interpret?
   _____

2. **Interpret Data** Where was wind fastest? Slowest? Which tells you this information more quickly, the table or the models?
   _____
   _____
   _____

3. **Communicate** Compare and contrast other weather conditions in the cities.
   _____
   _____
   _____
   _____

Name_____ Date_____

## Explore Activity
### Lesson 7

# How Can You Compare Weather?

**Hypothesize** How can you tell where the weather may change? Test your ideas. How would you use a weather map to give a weather report of the country? Write a **Hypothesis:**

_____

_____

_____

### Materials

- station model key
- newspaper weather map (optional)
- pencil
- crayons

## Procedure

**Communicate** Think of the country in large regions—the Northeast, the Southwest, and the coasts. Write a report for the weather in each region based on the map you see on page 152. Use another sheet of paper if necessary.

_____
_____
_____
_____
_____
_____
_____
_____
_____
_____
_____
_____

Name_____ Date_____

**Explore Activity**
**Lesson 7**

## Drawing Conclusions

1. **Infer** Which areas are having warm, rainy weather?

   _____

2. **Infer** Where is the weather cool and dry?

   _____

3. **Predict** How do you think weather in any part of the country may change, based on the data in this map? Give reasons for your answer. How would you check your predictions?

   _____

   _____

   _____

   _____

4. **FURTHER INQUIRY** **Interpret Data** What will tomorrow's weather be like? Interpret the information on the weather map in the morning paper. Compare your interpretation to the actual weather during the day.

   _____

**Inquiry**

Think of your own questions that you might like to test. Can temperatures vary within a small area?

My Question Is:

_____

_____

How I Can Test It:

_____

_____

My Results Are:

_____

Unit D · Astronomy, Weather and Climate          Use with textbook page D69

Name_____ Date_____

**Explore Activity**
**Lesson 7**

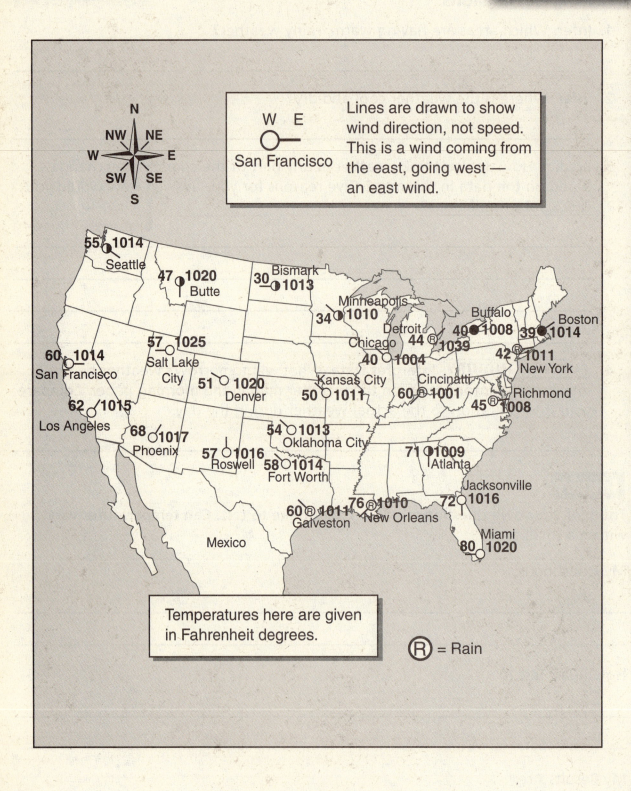

Unit D · Astronomy, Weather and Climate    Use with textbook page D69

Name_____ Date_____

**Alternative Explore**
Lesson 7

# On the Table

## Procedure

1. **Examine** Study the weather map on page 152 or a newspaper weather map. Think of the country in large regions—the Northeast, the Southwest, and so on. Think of regions like the Pacific Coast, the Atlantic Coast, and the Gulf Coast, in which you could group the weather data shown. List the regions you will use to group the weather data.

   _____
   _____
   _____

**Materials**
- weather map on page 152
- newspaper weather map (optional)
- station model key

2. **Organize** Record the data on the weather map in the table below. Be sure to group data from a single region together in the table. Continue the table on a separate sheet of paper.

| Region | City | Temperature (°F) | Air Pressure (mb) | Wind Direction | Cloud Cover |
|---|---|---|---|---|---|
|  |  |  |  |  |  |
|  |  |  |  |  |  |
|  |  |  |  |  |  |
|  |  |  |  |  |  |
|  |  |  |  |  |  |

## Drawing Conclusions

1. In which regions is the weather warm and rainy? Cold and dry?

   _____
   _____

2. How did the table help you to recognize regional weather patterns?

   _____
   _____

Unit D · Astronomy, Weather and Climate    Use with TE textbook page D69

Name_____ Date_____

# Weather Prediction

**Hypothesize** Can you predict what the weather will be?
Write a **Hypothesis**:
_____
_____

**Materials**
- newspaper

1. Find a weather map in a newspaper that shows the weather across the United States. Be sure the map shows at least one cold front or warm front in the western part of the country. Look at your map.

   _____

2. Describe the weather in each region of the country—northwest, southwest, southeast, northeast.

3. **Infer** Weather patterns move from west to east across the United States. How do you think the weather just east of the front will change in the next day or so?

   _____
   _____
   _____

Unit D · Astronomy, Weather and Climate          Use with textbook page D72

Name_____ Date_____

**QUICK LAB**
FOR SCHOOL OR HOME
Lesson 7

## Drawing Conclusions

4. **Going Further** Why do you think forecasting the weather is important?

_____

_____

_____

_____

Name _____ Date _____

# Where Do Tornadoes Occur?

**Explore Activity**
Lesson 8

**Hypothesize** Tornadoes strike all parts of the United States. However, they are more frequent in some regions than in others. Where in the U.S. is "tornado country"? How might you test your hypothesis? Write a **Hypothesis:**

_____
_____
_____

**Materials**
- blue marker
- red marker
- map of U.S., including Alaska and Hawaii, on page 158

| State | Total | Average per Year | State | Total | Average per Year | State | Total | Average per Year | State | Total | Average per Year | State | Total | Average per Year |
|---|---|---|---|---|---|---|---|---|---|---|---|---|---|---|
| AL | 668 | 22 | HI | 25 | 1 | MA | 89 | 3 | NM | 276 | 9 | SD | 864 | 29 |
| AK | 0 | 0 | ID | 80 | 3 | MI | 567 | 19 | NY | 169 | 6 | TN | 360 | 12 |
| AZ | 106 | 4 | IL | 798 | 27 | MN | 607 | 20 | NC | 435 | 15 | TX | 4,174 | 139 |
| AR | 596 | 20 | IN | 604 | 20 | MS | 775 | 26 | ND | 621 | 21 | UT | 58 | 2 |
| CA | 148 | 5 | IA | 1,079 | 36 | MO | 781 | 26 | OH | 463 | 15 | VT | 21 | 1 |
| CO | 781 | 26 | KS | 1,198 | 40 | MT | 175 | 6 | OK | 1,412 | 47 | VA | 188 | 6 |
| CT | 37 | 1 | KY | 296 | 10 | NE | 1,118 | 37 | OR | 34 | 1 | WA | 45 | 2 |
| DE | 31 | 1 | LA | 831 | 28 | NV | 41 | 1 | PA | 310 | 10 | WV | 69 | 2 |
| FL | 1,590 | 53 | ME | 50 | 2 | NH | 56 | 2 | RI | 7 | 0 | WI | 625 | 21 |
| GA | 615 | 21 | MD | 86 | 3 | NJ | 78 | 3 | SC | 307 | 10 | WY | 356 | 12 |

## Procedure

1. **Infer** The table shown here lists how many tornadoes occurred in each state over a 30-year period. It also shows about how many tornadoes occur in each state each year. Look at the data in the table for two minutes. Now write what part of the country you think gets the most tornadoes.

_____
_____

2. Use the red marker to record the number of tornadoes that occurred in each state over the 30-year period. Use the blue marker to record the average number of tornadoes that occurred in a year in each state.

Unit D · Astronomy, Weather and Climate        Use with textbook page D75

Name_____ Date_____

**Explore Activity**
**Lesson 8**

## Drawing Conclusions

1. **Use Numbers** Which states had fewer than 10 tornadoes a year? Which states had more than 20 tornadoes a year?

   _____
   _____
   _____
   _____

2. **Interpret Data** Which six states had the most tornadoes during the 30-year period?

   _____

3. **Interpret Data** Which part of the country had the most tornadoes?

   _____

4. **FURTHER INQUIRY** **Communicate** Many people refer to a certain part of the country as "Tornado Alley." Which part of the country do you think that is? Why do you think people call it that? What else might these states have in common? Describe how you would go about finding the answer to that question.

   _____
   _____
   _____
   _____
   _____
   _____

Unit D · Astronomy, Weather and Climate         Use with textbook page D75

Name_____ Date_____

# Explore Activity
## Lesson 8

**Inquiry**

Think of your own questions that you might like to test. Do tornadoes in "Tornado Alley" occur most often during certain times of the year?

My Question Is:
_____
_____
_____

How I Can Test It:
_____
_____
_____

My Results Are:
_____
_____

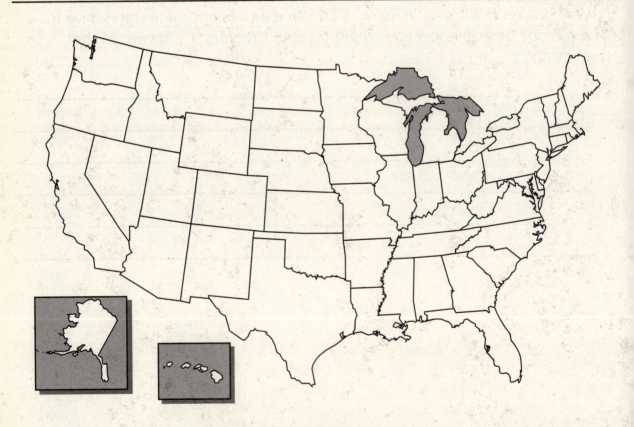

Name_____ Date_____

**Alternative Explore**
Lesson 8

# In the Alley

**Procedure**

1. Find out about the weather conditions that cause tornadoes to form. Use encyclopedias and any other research materials that may be helpful. Write down what you learned.

   _____
   _____
   _____

**Materials**
- encyclopedias and other research materials

2. Research the climate in the Central Plains region. What are conditions like there during the spring and early summer?

   _____
   _____
   _____

**Drawing Conclusions**

1. Why do you think tornadoes are common in the Central Plains region?

   _____
   _____
   _____

2. Compare the climate where you live to the climate in the Central Plains region. Are tornadoes common where you live? Why or why not?

   _____
   _____
   _____
   _____

Unit D · Astronomy, Weather and Climate      Use with textbook page D75

Name_____ Date_____

# Tornado in a Bottle

**Hypothesize** How does a tornado form?
Write a **Hypothesis**:
_____
_____
_____

**Materials**
- two 2-L plastic bottles
- duct tape
- water
- paper towel
- pencil

## Procedure

1. **Make a Model** Fill a 2-L plastic bottle one-third full of water. Dry the neck of the bottle, and tape over the top with duct tape. Use a pencil to poke a hole in the tape.

2. Place another 2-L plastic bottle upside down over the mouth of the first bottle. Tape the two bottles together.

3. **Observe** Hold the bottles by the necks so the one with the water is on top. Swirl them around while your partner gently squeezes on the empty bottle. Then place the bottles on a desk with the water bottle on top. Describe what you see.

_____
_____
_____

Name_____ Date_____

**Quick Lab FOR SCHOOL OR HOME — Lesson 8**

## Drawing Conclusions

4. **Infer** How is this like what happens when a tornado forms? Explain.

   _____
   _____

5. **Going Further** What kind of damage can tornadoes cause? Research newspaper articles and reference books to find out.

   My Hypothesis Is:

   _____
   _____

   My Experiment Is:

   _____
   _____

   My Results Are:

   _____
   _____
   _____

Name _____ Date _____

**Explore Activity**
Lesson 9

# What Do Weather Patterns Tell You?

**Hypothesize** What factors are used to describe the average weather pattern of a region? How might you use graphs of year-round weather in different places to test your ideas? Write a **Hypothesis:**

_____
_____
_____
_____

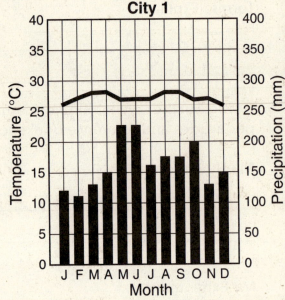

## Procedure

1. **Use Numbers** Look at the graph for city 1. The bottom is labeled with the months of the year. The left side is labeled with the temperature in degrees Celsius. Use this scale to read the temperature line. What is city 1's average temperature in July?

   _____
   _____

2. **Use Numbers** The right side of the graph is labeled millimeters of precipitation. Use this scale to read the precipitation bars. What is city 1's average precipitation in July?

   _____
   _____

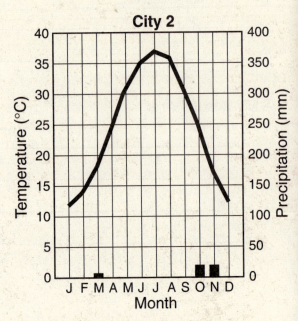

—— Temperature
■ Precipitation

3. Repeat steps 1 and 2 for city 2.

   _____

162    Unit D · Astronomy, Weather and Climate     Use with textbook page D83

Name _____ Date _____

**Explore Activity — Lesson 9**

## Drawing Conclusions

1. **Use Numbers** How do the annual amounts of precipitation compare for the two cities? Record your answer.

   _____

2. **Interpret Data** When is the average temperature highest for each city? Lowest? When does each city receive the greatest amount of precipitation?

   _____

   _____

   _____

   _____

3. **Interpret Data** Describe the average weather pattern for each city. Be sure to include temperature and precipitation, and their relationship to the seasons.

   _____

   _____

4. **FURTHER INQUIRY Communicate** What would a yearly graph for your community look like? Gather monthly temperature and precipitation data. Construct your graph. Compare it to city 1 and city 2.

   _____

### Inquiry

Think about your own questions that you might like to test. What causes the difference in weather conditions from place to place?

My Question Is:

_____

How I Can Test It:

_____

_____

My Results Are:

_____

Unit D · Astronomy, Weather and Climate        Use with textbook page D83

Name _____ Date _____

# Adapting to Climate Changes

## Procedure

1. Think about what might happen if you woke up one morning and found that the climate had drastically changed.

2. Write a story about it on a separate sheet of paper. Include how the climate changed and the adaptations you made in response to the climate changes.

## Drawing Conclusions

1. What adaptations did you make to the climate changes?

   _____
   _____
   _____

2. How was the climate different from before?

   _____
   _____
   _____

3. What are adaptations you make to regular climate changes during the year?

   _____
   _____
   _____
   _____

Name_____ Date_____

**Inquiry Skill Builder**
**Lesson 9**

# Measure

## Modeling Climates

In this activity you will make a model of the soil conditions in two cities. Use the information in the graphs on page D83. The soil conditions you set up will model—or represent—the climates of the two cities. To do this, you will need to measure the amount of water you use and the amount of time you use the lamp.

### Materials
- stick-on notepaper
- marking pencil or pen
- 2 trays of dry soil
- spray bottle of water
- lamp
- thermometer

## Procedure

1. **Measure** Put 3 cm of dry soil into each tray. Label one tray City 1 and the other tray City 2.
2. **Use Numbers** What do the bars on each graph represent? List the amounts given by the bars for each month for each city.

| Month | Precipitation (mm) City 1 | City 2 | Month | Precipitation (mm) City 1 | City 2 |
|---|---|---|---|---|---|
| January | | | July | | |
| February | | | August | | |
| March | | | September | | |
| April | | | October | | |
| May | | | November | | |
| June | | | December | | |

Unit D · Astronomy, Weather and Climate        Use with textbook page D85

Name _____ Date _____

**Inquiry Skill Builder**
**Lesson 9**

3. **Measure** Model the yearly precipitation and temperature like this: Let 5 minutes equal 1 month. One squeeze of water sprayed on the tray equals 10 millimeters of precipitation. Every minute the lamp is on equals 20 degrees of temperature. That means that from 0 to 5 minutes is January. During January the City 2 tray gets no water and the lamp shines on it for $\frac{3}{4}$ minute. The City 1 tray gets 12 squeezes of water and the lamp shines on it for $1\frac{1}{4}$ minutes.

4. **Make a Model** Model the two cities for all 12 months. Record your observations.

_____

_____

| Month | Observations: City 1 | Observations: City 2 |
|---|---|---|
| January | | |
| February | | |
| March | | |
| April | | |
| May | | |
| June | | |
| July | | |
| August | | |
| September | | |
| October | | |
| November | | |
| December | | |

**Drawing Conclusions**

1. **Observe** Examine the soil in the trays. Compare them for the same months. How do they differ?

_____

_____

2. **Communicate** How did measuring help you model climates?

_____

Name_____ Date_____

**Explore Activity**
**Lesson 1**

# Which Is More?

**Hypothesize** What properties do you use to compare the amounts of things? Are there different ways something can be "more" than other things? Write a **Hypothesis**:

_____
_____

**Materials**
- golf ball or wooden block
- blown-up balloon
- equal-pan balance
- ruler
- string
- box, such as a shoe box, big enough for the balloon to fit in
- pail of water

## Procedure: Design Your Own

1. **Observe** Look at the golf ball (or wooden block) and blown-up balloon. Which is "more"? Think of how one object could be "more":
   - more when you use a balance
   - more when you put it in water and see how much the water level goes up, and so on
   
   Record your observations.
   
   _____
   _____

2. Use the equipment to verify one way that one object is more than another. Decide which of the two objects is "more" and which one is "less."
   
   _____
   _____

3. Repeat your measurements to verify your answer.
   
   _____

4. Now use different equipment to compare the two objects. Is the same object still "more"? Explain.
   
   _____
   _____

5. Repeat your measurements to verify your answer.
   
   _____

Name_____ Date_____

## Drawing Conclusions

1. **Communicate** Identify the equipment you used. Report your results.

   _____
   _____
   _____

2. For each test, which object was more? In what way was it more than the other object?

   _____
   _____
   _____
   _____

3. **FURTHER INQUIRY** **Experiment** What if you were given a large box of puffed oats and a small box of oatmeal? Which do you think would be more? Design an experiment to test your hypothesis. Tell what equipment you would use.

   _____
   _____
   _____
   _____
   _____

**Inquiry**

Think of your own questions that you might test. How might you compare two objects that are similar in size?

My Question Is:

_____

How I Can Test It:

_____

My Results Are:

_____

Unit E · Properties of Matter and Energy       Use with textbook page E5

Name_____  Date_____

# Which Bean Is More?

**Procedure**

1. Count the number of jelly beans needed to fill one cup. Record the number in the table.
2. Repeat step 1 using dry beans.
3. Place each full cup on opposite sides of an equal-pan balance. Observe which cup has more mass and record your observation in the table.
4. Use a ruler to measure the length of a jelly bean. Record the length in the table.
5. Repeat step 4 using a dry bean.

**Materials**
- small foam cups
- jelly beans
- dry beans
- equal-pan balance
- ruler

|  | Jelly Beans | Dry Beans |
|---|---|---|
| Number to fill cup |  |  |
| Which has more mass? |  |  |
| Length |  |  |

**Drawing Conclusions**

1. Which bean took more to fill the cup? Why?

   _____
   _____
   _____

2. Which cup of beans had more mass? How could you tell?

   _____
   _____

3. Which bean was longer?

   _____

| Name | Date |

## Inquiry Skill Builder
### Lesson 1

# Make a Model

### How Metal Boats Float

Think about objects that have more matter packed into the space they take up than water does. Will such objects sink or float in water? You have probably seen how a metal object like a nail or a spoon sinks in water. However, huge ships made of similar metal float even when they carry large cargoes. How is this possible? In this activity, you will make different sized models of a metal boat. Scientists use models to help them understand properties of matter. Models also makes experimenting easier. Try different designs to see how well the model boats can carry heavy cargo.

### Materials
- aluminum foil
- large paper clips
- pan of water

### Procedure

1. **Make a Model** Prepare 3 sheets of aluminum foil of different sizes. Record their lengths and widths and use them to make 3 boats. Experiment with different designs and float them on water.

2. **Predict** Write down what you think will happen when you place more and more matter in the empty space of the boat. What steps should you follow to test your prediction? Be sure to use only the materials listed above.

   _____
   _____

3. **Experiment** Carry out your procedure, keeping a written record of what you observe.

   _____
   _____
   _____
   _____
   _____

Name _____ Date _____

**Inquiry Skill Builder**
**Lesson 1**

## Drawing Conclusions

1. **Communicate** How well did your results agree with your prediction?
   _____
   _____

2. Compare your model with those of your classmates. Which boat held the most clips? Why?
   _____
   _____
   _____

3. **Make a Model** The aluminum foil boat is a model of a steel ship. Use the way your boat floats to explain how a steel ship floats. Why was using a model of a large ship helpful?
   _____
   _____
   _____

4. **Communicate** What changed as more and more matter was added to the empty space of the boat? What happened as a result of this change?
   _____
   _____
   _____
   _____
   _____
   _____

Unit E · Properties of Matter and Energy          Use with textbook page E9

Name_____ Date_____

**Explore Activity**
Lesson 2

# How Do We Know What's "Inside" Matter?

**Materials**
- 3 identical, sealed, opaque boxes
- equal-pan balance with set of masses
- magnet

**Hypothesize** How can you tell what is inside a sealed opaque box without opening it? What sorts of tests would you perform to try to identify its contents?
Write a **Hypothesis:**
_____
_____

## Procedure

1. **Observe** Examine the three boxes, but do not open them. You can lift them, shake them, listen to the noises they make, feel the way their contents shift as you move them, and so on. Use the magnet and balance to obtain more data.
_____
_____

2. **Infer** Try to determine what is in each box.
_____
_____

| Box | Observations | Object(s) |
|-----|--------------|-----------|
| #1  |              |           |
| #2  |              |           |
| #3  |              |           |

## Drawing Conclusions

1. **Communicate** Describe what you think is in each box.
_____
_____

Unit E · Properties of Matter and Energy — Use with textbook page E21

Name_____ Date_____

2. How did you make your decision?
   _____

3. Do these boxes have anything in common? In what ways are they similar? In what ways are they different?
   _____
   _____

4. **FURTHER INQUIRY** Experiment What if you have a can of peanuts and a can of stewed tomatoes? The cans look the same except for the labels. Now what if your baby brother takes the labels off? You want the peanuts, but you don't want to open the tomatoes by mistake. What experiments can you do to find out what is inside—before you open the cans?
   _____
   _____

**Inquiry**

Think of your own questions that you might test. How might I tell the difference between a liquid and a gas?

My Question Is:

_____

How I Can Test It:

_____

My Results Are:

_____
_____

Name_____ Date_____

# What's in Me?

## Procedure

1. Form a group with several other students.

2. **Experiment** You must try to determine the contents of each of the boxes. Start by discussing what tests you will conduct to make observations that will help you infer the contents.

3. **Observe** Conduct your tests to determine the contents of each of the boxes. After each test, record your observations in the table.

**Materials**
- 2 sealed boxes of different sizes
- soft toy
- colored pencils or crayons

| Test | OBSERVATIONS | |
|---|---|---|
| | Smaller Box | Larger Box |
| 1 | | |
| 2 | | |

4. **Infer** Try to determine what is in each box.

## Drawing Conclusions

1. **Communicate** In the space provided, make a drawing of what you think is inside each box.

Smaller Box                                   Larger Box

2. How did you make your decisions?

_____

_____

Unit E · Properties of Matter and Energy        Use with TE textbook page E21        175

Name _____ Date _____

# Modeling Molecules

**Hypothesize** How do different elements combine to form molecules? Write a **Hypothesis:**

_____
_____
_____
_____

**Materials**
- large and small marshmallows
- toothpicks

## Procedure

1. Using small marshmallows for hydrogen atoms and large marshmallows for oxygen atoms, make two $H_2$ molecules and one $O_2$ molecule. Join the atoms with toothpicks.

2. **Use Numbers** Count the number of atoms of each type you have in your molecules. Record these numbers. Take these same marshmallows and make as many water molecules as you can, using toothpicks to join the atoms.

_____
_____

## Drawing Conclusions

3. **Observe** How many water molecules did you make?

_____

4. **Infer** Why would real water molecules have properties different from real hydrogen and oxygen molecules?

_____
_____

Name_____ Date_____

5. **Going Further** Describe other examples of elements combining to form new compounds. Write and conduct an experiment.

My Hypothesis Is:

_____

_____

My Experiment Is:

_____

_____

_____

My Results Are:

_____

_____

Name_____ Date_____

# What Happens When Ice Melts?

**Hypothesize** How does the temperature change as a block of ice melts? Does it increase?

_____
_____
_____

### Materials
- ice cubes
- water
- graduated cylinder
- plastic or paper cup
- thermometer
- heat source (lamp or sunlight)
- watch or clock
- equal-pan balance with set of masses

## Procedure

1. **Predict** You are going to explore the effect of heat on ice cubes as they melt. How do you think the temperature and mass of a cup of ice will change as it melts? Make a graph showing your prediction of how the temperature will change.

   _____
   _____

2. **Measure** Put ice cubes in the cup. Add 50mL of water. Find and record the mass of the mixture. Swirl the mixture for 15 seconds.

3. **Measure** Place the thermometer in the cup. Wait 15 seconds. Then read and record the temperature. Put the cup under a heat source (lamp or sunlight). Take temperature readings every 3 minutes. Record each result.

   _____
   _____

4. **Measure** After all the ice has melted, find and record the mass of the cup of water again. Take and record 5 more water temperature readings at 3-minute intervals.

   _____

| Time | Temperature (As Ice Melts) | Temperature (After Ice Melts) |
|---|---|---|
| 3 minutes |  |  |
| 6 minutes |  |  |
| 9 minutes |  |  |
| 12 minutes |  |  |
| 15 minutes |  |  |

Name_____ Date_____

## Drawing Conclusions

1. **Observe** What happened to the mass and temperature?
   _____
   _____

2. **Hypothesize** Why do you think you got the results described in question 1?
   _____
   _____
   _____

3. **FURTHER INQUIRY** **Predict** Design an experiment to test each of your predictions. What do you think will happen as you freeze water? Design an experiment to test your prediction.
   _____
   _____
   _____

### Inquiry

Think of your own questions that you might test. How might the intensity of heat affect melting ice?

My Question Is:
_____
_____

How I Can Test It:
_____
_____

My Results Are:
_____

Unit E · Properties of Matter and Energy    Use with textbook page E35

Name_____ Date_____

## Alternative Explore
### Lesson 3

# Melting Temperature

## Procedure

**BE CAREFUL!** Do not touch the heated surface of the hot plate.

1. Form a group with several other students.
2. Select three samples from the set of melting samples.
3. Place the items on the aluminum foil or pie plate. Then place them on the hot plate.
4. **Observe** Note the order in which the items melt.

_____
_____
_____

### Materials
- oven mitt
- goggles
- hot plate
- aluminum foil or pie plate
- melting samples (ice, wax, butter, chocolate, and mozzarella cheese)

## Drawing Conclusions

1. **Interpret Data** Organize the data by placing the item that melted last at the top of the list and the item that melted first at the bottom.

   **Order of Melting**
   _____

2. **Hypothesize** Why do you think you got the results described in your list?
   _____
   _____

Name_____ Date_____

# Collapsing Bottles

**QUICK LAB**
FOR SCHOOL OR HOME
Lesson 3

**Hypothesize** How does heat affect an empty plastic bottle? How does cold affect it? Write a **Hypothesis:**

_____
_____

### Materials

- flexible plastic bottle with screw cap
- pails of hot and ice-cold water

## Procedure

1. **Predict** How does heat affect an empty plastic bottle? What do you think will happen to a bottle when it is warmed? What do you think will happen to it when it is cooled? Record your predictions.

   _____
   _____

2. With the cap off, hold a bottle for a minute or two in a pail of hot tap water. Then screw the cap on tightly while the bottle is still sitting in the hot water.

3. **Experiment** Now hold the bottle in a pail of ice water for a few minutes.

## Drawing Conclusions

4. **Communicate** Record your observations.

   _____
   _____

5. **Infer** Write out an explanation of why the bottle changed as it did. Be sure to use the idea of how molecules move at different temperatures.

   _____
   _____
   _____
   _____

Unit E · Properties of Matter and Energy  Use with textbook page E41

Name_____ Date_____

6. **Going Further** Do gases also contract and expand with changing temperatures? Write and conduct an experiment.

My Hypothesis Is:

_____
_____

My Experiment Is:

_____
_____
_____
_____
_____

My Results Are:

_____
_____
_____
_____
_____
_____

Name_____ Date_____

**Explore Activity**
Lesson 4

# How Can You Take Apart Things that Are Mixed Together?

**Hypothesize** How can you separate substances that are mixed together in a way that they keep their properties?
Write a **Hypothesis:**

_____
_____
_____

### Materials
- sample of substances mixed together
- hand lens
- toothpicks
- magnet
- paper (coffee) filters
- 2 cups or beakers
- water
- goggles

## Procedure: Design Your Own

**BE CAREFUL!** Wear goggles. Do not taste your sample.

1. **Observe** Examine the sample your teacher gives you. It is made of different substances. One of the substances is table salt. What else does it seem to be made of? Record your observations.

_____
_____
_____

2. **Experiment** Design and carry out an experiment to separate the various ingredients in your sample.

_____
_____

| Sample | Observations |
|---|---|
| Color | |
| Size | |
| | |
| | |

Unit E · Properties of Matter and Energy — Use with textbook page E51 — **183**

Name _____ Date _____

## Drawing Conclusions

1. **Infer** How many parts or substances were mixed into your sample? How did you reach that conclusion?

   _____

2. You knew one substance was salt. What properties of salt might help you separate it from the rest? Could you separate salt first? Why or why not?

   _____
   _____
   _____

3. How did you separate out the substances? How did you use the properties of these substances to separate them?

   _____
   _____
   _____

4. **FURTHER INQUIRY** **Experiment** What if you were given white sand and sugar mixed together? How would you separate the two ingredients?

   _____
   _____
   _____

### Inquiry

Think of your own questions that you might test. Would magnetism help you separate iron filings from solids?

My Question Is:

_____

How I Can Test It:

_____

My Results Are:

_____

Name_____ Date_____

**Alternative Explore**
**Lesson 4**

# Separate by Dissolving

**Procedure** BE CAREFUL! Wear goggles.

1. Prepare a mixture by placing a spoonful each of salt and ground wax into a plastic glass. Stir the mixture.
2. Add a little water to the glass.
3. Stir the contents until the salt dissolves. (Add more water if you need to.)
4. Tuck a coffee filter into a second plastic glass.
5. **Observe** Pour the mixture into the filter. Record your observations.

_____
_____

6. Remove the filter. Place the glass of liquid in a sunny spot.
7. **Observe** After several days, observe the contents of the glass and record your observations.

_____
_____

**Materials**
- salt
- ground wax
- goggles
- water
- coffee filter
- 2 plastic glasses
- spoon

## Drawing Conclusions

1. At what point did you separate the wax from the salt? Explain your answer.

_____
_____

2. **Identify** What property did you use to separate the mixture? Explain your answer.

_____
_____

Unit E · Properties of Matter and Energy    Use with TE textbook page E51

Name_____ Date_____

# Solubility

**Hypothesize** Is a substance more soluble in one piece or in many pieces? Test your ideas. Write a **Hypothesis:**

_____

_____

### Materials
- water
- 3 cups
- 100 mL beaker
- sugar cubes

## Procedure

1. **Measure** Add 100 mL of water to each of 3 cups. Place a sugar cube in one of the cups, and stir until the cube is dissolved. Record the time it took to dissolve.

   _____

2. **Predict** Carefully break a second sugar cube into two pieces. How long do you think it will take to dissolve? Place the two pieces in a second cup of water, and stir until they dissolve. Record the time they took to dissolve.

   _____

   _____

3. **Use Variables** Fold a piece of paper around a sugar cube, and carefully break the cube into many pieces. Pour the pieces into the third cup of water, and stir until dissolved. Record the dissolving time.

   _____

Name_____ Date_____

**QUICK LAB FOR SCHOOL OR HOME**
Lesson 4

## Drawing Conclusions

4. **Interpret Data** Construct a graph that illustrates your findings. Which sugar cube dissolved the fastest?

   _____
   _____

5. What conclusion can you make regarding dissolving time based on the experiment?

   _____
   _____

6. **Going Further** Think of your own question that you might like to test. Do different substances dissolve at the same rates?

   My Hypothesis Is:

   _____
   _____

   My Experiment Is:

   _____
   _____
   _____

   My Results Are:

   _____
   _____

Unit E · Properties of Matter and Energy  Use with textbook page E58

Name_____ Date_____

**FOR SCHOOL OR HOME**
**Lesson 4**

# Kitchen Colloids

**Hypothesize** What happens to cream when you whip it?
Write a **Hypothesis:**

_____

_____

### Materials
- whipping cream
- 2 bowls
- wire whisk
- ice

## Procedure

1. Pour some whipping cream into a bowl. Set it in a bed of ice in another bowl. Let the cream and bowl chill. Whip the cream until it is fluffy.

2. **Observe** Let the cream warm. Whip it more. How does it change? Record your observations.

   _____

   _____

## Drawing Conclusions

3. **Interpret Data** What kind of colloids did you make in steps 1 and 2?

   _____

   _____

4. **Infer** What are colloids commonly known as?

   _____

   _____

Name_____ Date_____

5. **Going Further** What other colloids can you find in the kitchen? What are they made of? Write and conduct an experiment.

My Hypothesis Is:
_____

My Experiment Is:
_____
_____

My Results Are:
_____
_____
_____

Name_____ Date_____

# How Can You Recognize a Chemical Change?

**Hypothesize** How can you tell if a substance changes into something else? What signs would you look for?

Write a **Hypothesis:**

_____
_____
_____

**Materials**
- baking soda
- baking powder
- cornstarch
- salt
- iodine solution
- vinegar
- water
- wax paper
- permanent marker
- 4 toothpicks
- 3 droppers
- 4 plastic spoons
- 7 small cups
- goggles

**Procedure** **BE CAREFUL!** Wear goggles.

1. Copy the grid on page 192 on wax paper with a marking pen. Using a spoon, put a pea-sized amount of cornstarch in each of the three boxes in the first row.

2. **Observe** Use a dropper to add five drops of water to the cornstarch in the first column. Stir with a toothpick. Record your observations.

_____
_____

3. **Experiment** Using a different dropper, add five drops of vinegar to the cornstarch in the second column. Stir with a new toothpick. Record your observations.

_____

4. **Observe** Use a third dropper to add five drops of iodine solution to the cornstarch in the third column. Record your observations. CAUTION: Iodine can stain and is poisonous.

_____
_____

Name_____ Date_____

**Explore Activity**
**Lesson 5**

5. **Experiment** Repeat steps 1–4 for baking powder, baking soda, and salt.

_____
_____
_____
_____
_____

## Drawing Conclusions

1. **Infer** In which boxes of the grid do you think substances changed into new substances? Explain your answers.

_____
_____
_____

2. **FURTHER INQUIRY** **Infer** Your teacher will give you samples of two unknown powders. Use what you have learned to identify these powders. Report on your findings.

_____
_____
_____
_____
_____

Name_____ Date_____

**Explore Activity**
Lesson 5

**Inquiry**

Think of your own questions that you might test. What evidence might show a liquid changing to a gas?

My Question Is:

_____

_____

How I Can Test It:

_____

My Results Are:

_____

|  | Water | Vinegar | Iodine solution |
|---|---|---|---|
| Cornstarch |  |  |  |
| Baking powder |  |  |  |
| Baking soda |  |  |  |
| Salt |  |  |  |

Name _____ Date _____

**Alternative Explore**
**Lesson 5**

# Form Rust

## Procedure

1. Wet the inside of the test tube by filling it with water. Then pour out the water.
2. Place a small amount of steel wool in the bottom of the test tube.
3. Fill the beaker with water.
4. Place your finger over the opening of the test tube. Turn the test tube upside down. Continuing to block the opening with your finger, place the test tube in the beaker of water.
5. Let go of the test tube and place the beaker where it will not be disturbed.
6. **Observe** Examine the test tube over the next few days. Record your observations.

**Materials**
- steel wool (iron)
- water
- test tube
- beaker

_____
_____
_____

## Drawing Conclusions

1. **Infer** What evidence did you observe that a chemical change took place?
_____
_____

2. What substances do you think reacted to produce the chemical change?
_____
_____

Unit E · Properties of Matter and Energy          Use with TE textbook page E69

**Name** _____ **Date** _____

# Experiment

**Inquiry Skill Builder**
**Lesson 5**

## Preventing Rust

You've learned that steel forms rust when exposed to oxygen and moisture. Rusting can ruin metal objects. Can you find a way to stop or slow rusting? In this activity you will experiment to try to find the answer. In order to experiment, you need to do the following things. Form a hypothesis. Design a control. Carry out your experiment. Analyze and communicate your results.

**Materials**
- steel nails and sandpaper
- paper cups
- dilute salt water
- goggles

**Procedure** **BE CAREFUL!** Wear goggles.

1. **Hypothesize** You can make a steel nail rust by placing it in water. Think of a way to protect a steel nail from rusting under such conditions. Write an explanation of why you think your method will work.

   _____
   _____

2. **Experiment** To test your method of rust protection, you need a control nail kept under normal conditions. Each experimental nail will have just one condition (variable) change. For example, what if you wanted to make a nail rust? You might leave one nail in a clean, empty jar (the control). You might put another in water. You might put a third in lemon juice. The amount of rusting that occurs is called the *dependent variable*. Write out how you will set up the experimental and control the nails for your experiment.

   _____
   _____
   _____
   _____

Unit E · Properties of Matter and Energy

Use with textbook page E75

**Inquiry Skill Builder**
Lesson 5

3. **Experiment** Carry out your experiment, and record your observations. Draw how the nails looked at the end of your experiment.

_____

_____

_____

## Drawing Conclusions

1. **Infer** Write out a description of how well your hypothesis agreed with your results. Be sure to compare the experimental nail with the control nail.

_____

_____

_____

2. **Communicate** Why did you need a control in this experiment?

_____

_____

_____

Name_____ Date_____

**Explore Activity**
Lesson 6

# Which Are Acids and Which Are Bases?

**Hypothesize** Can you test whether or not a solution is an acid or a base? Test your ideas. Write a **Hypothesis**:

_____

_____

**Procedure** BE CAREFUL! Wear goggles, gloves, and an apron.

1. **Predict** Which solutions do you think are acids and which are bases? Write your predictions in a chart like the one shown below.

2. **Observe** Vinegar is an acid. Put a small amount in a cup, and mark the cup with a label. Test by dipping a piece of red litmus paper into the vinegar. Record the result in your table. Repeat with a piece of blue litmus paper. Litmus paper is a material that allows you to tell which solutions are acids and which are bases.

3. **Classify** Test all of your other solutions in the same way, and record your results.

**Materials**

- red and blue litmus paper
- wide-range pH paper
- plastic cups
- labels
- goggles
- gloves
- apron
- household solutions

| Sample | Predict: Acid or Base? | Effect on Red Litmus | Effect on Blue Litmus | Result: Acid or Base? |
|---|---|---|---|---|
| Vinegar | ACID | | | ACID |
| Baking soda | | | | |
| Lemon juice | | | | |

**Drawing Conclusions**

1. Which samples are acids? How do you know?

_____

_____

196  Unit E · Properties of Matter and Energy   Use with textbook page E81

Name_____ Date_____

**Explore Activity**
**Lesson 6**

2. Which samples are bases? How do you know?

   _____
   _____

3. **Measure** Now test each sample with a small strip of pH paper. Match the color of the paper to the color scale on the holder, and find the pH.

   _____
   _____

4. **FURTHER INQUIRY** Interpret Data Design and do an activity to test the acidity of the foods you eat. Which foods are acidic? Which are basic? How do you know?

   _____
   _____
   _____

## Inquiry

Think of your own question that you might like to test. Could you test whether a solution is an acid or a base if you only had red litmus paper?

My Question Is:

_____
_____

How I Can Test It:

_____
_____

My Results Are:

_____
_____

Name_____ Date_____

Lesson 6

# Cleaning Pennies

## Procedure

1. Work in a group. Place one penny in the cup of water and baking soda solution. Place one penny in the cup of vinegar.

2. Swish the pennies around in both cups.

3. Take the pennies out of the cups. Record the results.

   _____
   _____
   _____

## Materials

- water and baking soda solution
- vinegar
- 2 paper cups for each group
- enough dirty pennies to provide 2 pennies per group of students

## Drawing Conclusions

1. Does vinegar or baking soda clean pennies better?

   _____

2. What other solutions might give similar results?

   _____
   _____

Name_____ Date_____

# Mystery Writing with a Base

**Hypothesize** Can grape juice be used to indicate the presence of a base substance? Test your ideas. Write a **Hypothesis**:

_____
_____

## Materials

- baking soda
- grape juice
- cotton swabs
- white drawing paper

## Procedure

1. Dip a cotton swab in baking soda solution. Use it to write a short message to your partner.

2. Put the paper aside, and allow it to completely dry. After it is dry, give it to your partner.

3. Can you read the message your partner gave you? You probably cannot. Use another cotton swab and gently "paint" the paper with grape juice.

4. **Observe** What happened when you painted the paper with the grape juice?

_____
_____
_____

Unit E · Properties of Matter and Energy    Use with textbook page E85

Name_____ Date_____

**QUICK LAB**
FOR SCHOOL OR HOME
Lesson 6

## Drawing Conclusions

5. **Infer** Is the grape juice an indicator? Why or why not?

   _____
   _____

6. **Going Further** Think of your own question that you might like to test. Would the grape juice work as an indicator if you used an acidic solution instead of a basic solution?

   My Hypothesis Is:

   _____
   _____

   My Experiment Is:

   _____
   _____

   My Results Are:

   _____
   _____
   _____
   _____

Unit E · Properties of Matter and Energy                Use with textbook page E85

Name_____ Date_____

**Explore Activity**
Lesson 7

# How Well Do Batteries Provide Energy?

**Materials**

For each circuit to be tested:
- battery
- flashlight bulb
- 2 wires

**Hypothesize** Is it better to buy heavy-duty batteries or less expensive ones? Which last longer? Which ones are really the least expensive to use? Write a **Hypothesis:**

_____
_____

## Procedure

1. **Experiment** In this activity you will determine which battery is the best buy. Test variables such as battery type, size, voltage, brand, or cost. Connect the wires to the battery and bulb as shown. Fasten the wires with a battery holder or tape. Record the time the bulb went on and the type, size, voltage, and brand of battery used. Share your data with the class.

2. **Observe** Check the bulb every 15 minutes to see if it is still lit. Record the time it goes off.

3. Repeat using another variable.

| Time | Observations |
|---|---|
| 15 minutes | |
| 30 minutes | |
| 45 minutes | |
| 1 hour | |
| 1 hr. 15 min. | |
| 1 hr. 30 min. | |

Unit E · Properties of Matter and Energy — Use with textbook page E91

Name_____ Date_____

## Drawing Conclusions

1. **Use Numbers** Divide the time each battery lasted by its cost.
   _____
   _____

2. Make a graph of the class's results. Which batteries lasted the longest? Which batteries cost the least per hour of use?
   _____
   _____
   _____

3. **Infer** Which batteries are the best buy? the cheapest? the longest lasting?
   _____
   _____
   _____
   _____

4. **FURTHER INQUIRY** **Interpret Data** Design an experiment to see if a battery will only last half as long if two bulbs are connected to it compared to only one. Does it matter how the bulbs are connected?
   _____
   _____
   _____
   _____

### Inquiry

Think of your own questions that you might test. What other factors affect the cost effectiveness of batteries?

My Question Is:

_____

How I Can Test It:

_____

My Results Are:

_____

Name_____ Date_____

**Alternative Explore**
Lesson 7

# Flashlight Test

**Procedure**

**Materials**
- flashlight with new batteries

1. Turn on the flashlight. Place the flashlight so you can easily observe its light.

2. **Observe** Record the time and whether the flashlight is lit in the table.

3. **Observe** Every 15 minutes throughout the day, repeat step 2 until the flashlight goes out. (If needed, continue the table on separate piece of paper.)

| Time | Is flashlight still lit? |
|------|--------------------------|
|      |                          |
|      |                          |
|      |                          |
|      |                          |
|      |                          |

**Drawing Conclusions**

1. How long did the flashlight stay lit?

   _____

2. **Compare and Contrast** Find the cost of the flashlight batteries from your teacher, then calculate the cost per hour of using the flashlight. Show your calculations. Compare your results with those of others.

   _____

   _____

   _____

Unit E · Properties of Matter and Energy        Use with TE textbook page E91

Name_____   Date_____

# Measure Electricity

**Hypothesize** Can electricity affect a magnet? Can a magnet be used to measure electricity? Write a **Hypothesis:**

_____
_____
_____

**Materials**
- compass
- 5 m of fine varnish-coated wire
- sandpaper
- 1.5-V battery and bulb circuit

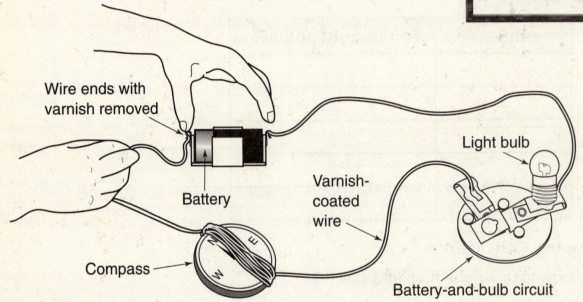

## Procedure

1. Wrap fine varnished wire around a compass. Remove the coating from the ends of the wire with sandpaper.

2. Turn the compass until the needle is lined up with the coils of wire.

3. Keeping the compass this way, connect the ends of the wire to a circuit of a battery and small light bulb. See the diagram.

4. **Observe** What sight tells you that electricity is flowing in the circuit?

_____

Name_____ Date_____

## Drawing Conclusions

**QUICK LAB**
**FOR SCHOOL OR HOME**
**Lesson 7**

5. **Observe** What does the needle do as you open and close the circuit?

   _____
   _____
   _____

6. **Infer** How might a more powerful battery affect the needle?

   _____
   _____

7. **Infer** How could the compass needle be used to measure electricity?

   _____
   _____

8. **Going Further** How do some animals use electric fields? Write and conduct an experiment.

   My Hypothesis Is:

   _____
   _____

   My Experiment Is:

   _____
   _____

   My Results Are:

   _____
   _____
   _____

Unit E · Properties of Matter and Energy     Use with textbook page E93

Name_____ Date_____

# How Fast Does a Spring Move Objects?

**Hypothesize** Will changing the amount of mass affect the swinging of a hanging mass? Test your ideas.

Write a **Hypothesis:**

_____
_____

**Procedure** BE CAREFUL! Wear goggles.

### Materials
- 3 masses (washers or AAA batteries)
- metal ruler 30.5 cm (12 in)
- rubber bands
- clock with second hand
- graph paper
- goggles

1. Attach a mass to the end of a metal ruler with a rubber band. Hold the ruler tightly against the edge of a table as shown so it can act like a spring.

2. **Use Numbers** Pull the mass back 5 cm (2 in.), and release it crisply. Count and record how many swings the mass completes in ten seconds.

_____
_____

3. **Predict** How will adding more mass to the end of the ruler affect how fast it swings back and forth? Record your predictions.

_____
_____

4. Add a second mass, and repeat the procedure. Repeat again with a third mass.

## Drawing Conclusions

1. **Infer** Why does the ruler move the attached mass when it is pulled back and released?

_____
_____

206     Unit F · Motion and Energy     Use with textbook page F5

Name_____ Date_____

**Explore Activity**
**Lesson 1**

2. **Observe** What effect did increasing the mass have on how fast the mass was swung back and forth by the ruler?

_____
_____

3. **Hypothesize** Why do you think the increase in mass had this effect?

_____
_____

4. **Interpret Data** Make a graph of your results. What two variables should you plot?

_____
_____

5. **FURTHER INQUIRY** **Predict** Use your graph to estimate how many swings would be observed in ten seconds when four masses are attached to the ruler.

_____
_____

## Inquiry

Think of your own question that you might like to test. How would the length of a ruler affect its swing?

My Question Is:

_____
_____

How I Can Test It:

_____
_____

My Results Are:

_____
_____

Unit F · Motion and Energy            Use with textbook page F17

Name_____ Date_____

**Alternative Explore**
**Lesson 1**

# Measuring Mass

## Procedure

1. Attach a rubber band to an empty laboratory cart. Stretch the rubber band back as far as possible.
2. Now release the cart. Observe its speed.
3. Repeat the experiment, this time loading the cart with weights.
4. Compare and contrast the speed of the empty and the loaded carts.

### Materials
- laboratory carts
- weights
- rubber bands

## Drawing Conclusions

1. Which cart moved faster? Why?

   _____
   _____

2. What might happen if you repeated the experiment, filling the first cart with rocks and the second with feathers? Explain the results.

   _____
   _____
   _____

Name_____  Date_____

# Using a Position Grid

**Hypothesize** How can you show an object's position relative to other objects? Test your ideas. Write a **Hypothesis:**
_____
_____

**Materials**
- graph paper

## Procedure

1. A grid has rows and columns. Each is labeled with letters or numbers. You can locate each box in the grid by its letter and number address. See the map grid for examples. Make your own grid using graph paper. Number the boxes from 1 to 29 across and from A to G down. Explain why the snail didn't want to move. How is a position grid useful?

|   | 1 | 2 | 3 | 4 | 5 | 6 | 7 | 8 | 9 | 10 |
|---|---|---|---|---|---|---|---|---|---|----|
| A |   |   |   |   |   |   |   |   |   |    |
| B |   |   |   |   |   |   |   |   |   |    |
| C |   |   |   |   |   |   |   |   |   |    |
| D |   |   |   |   |   |   |   |   |   |    |
| E |   |   |   |   |   |   |   |   |   |    |
| F |   |   |   |   |   |   |   |   |   |    |
| G |   |   |   |   |   |   |   |   |   |    |

2. Find each box, and shade it in with a colored pencil.

E27, B8, F24, D15, B29, C20, D5, F14, D29, B3, D11, B16, F3, D7, B27, B2, B11, F27, B20, F12, B23, D8, E17, F20, F23, C6, E2, E24, B10, B1, F10, C29, C2, F17, D24, E8, B15, E14, F1, B12, F5, D16, B21, B24, D27, C5, C10, E29, E7, B5, C14, C24, C16, E20, D2, C27, D10, C8, D17, E28, E10, D6, F25, D20, D14, F8, F29, B19, B14, F11, E5, B25, F2, B28

Unit F · Motion and Energy     Use with textbook page F10

Name_____ Date_____

# Drawing Conclusions

**QUICK LAB FOR SCHOOL OR HOME — Lesson 1**

3. Why did the snail not want to move?

   _____

4. **Infer** How is the grid that spelled out the reason for the snail not moving like the streets shown on a map of a city? How are street addresses like the letters and numbers that labeled the grid?

   _____
   _____
   _____
   _____

5. **Going Further: Predict** How might the answer change if you reversed the numbers and went from 29 to 1 from left to right?

   _____
   _____
   _____
   _____

Name_____ Date_____

# How Do Different Forces Affect an Object's Motion?

**Hypothesize** Will changing the force applied to an object affect its motion? Test your ideas.

Write a **Hypothesis:**

_____

_____

**Materials**
- toy car
- 2 boards with hooks for rubber bands
- rubber bands
- meterstick
- masking tape
- goggles
- stop watch
- compass

**Procedure** **BE CAREFUL!** Wear goggles.

1. Place a 15-cm (6-in.) strip of masking tape on the floor. Hold two boards on either side of the tape with a rubber band stretched between them.

2. **Measure** Pull a toy car back 5-cm (2-in.) against the rubber band to launch it. Use the compass, stopwatch and meterstick to determine the car's direction, elapsed time, and distance of travel.

3. **Observe** Repeat step 2 twice more. Record your results. Find the average speed.

4. **Predict** What will happen if you use two or three rubber bands to launch the car? Test your prediction.

**Drawing Conclusions**

1. **Interpret Data** When did the car move farthest on average—when one, two or three rubber bands were used?

_____

_____

_____

Unit F · Motion and Energy    Use with textbook page F17

Name_____ Date_____

2. **Infer** How is the distance traveled by the car in any trial related to the speed it was given by the rubber band? Why?

   _____
   _____

3. **FURTHER INQUIRY** **Predict** If you taped a second toy car on top of the first and launched them with two rubber bands, how far would the cars travel? Test your prediction. Record and explain your observations.

   _____
   _____
   _____

### Inquiry

Think of your own question that you might like to test. What would happen if you changed the surface that the cars were rolling on?

My Question Is:

_____
_____

How I Can Test It:

_____
_____

My Results Are:

_____
_____

Name_____ Date_____

**Alternative Explore**
**Lesson 2**

# Accelerating Masses

## Procedure

1. Throw each of the various balls as far as you can. Note that if you throw the balls at a 45-degree angle with the ground, you can achieve the greatest distance.

2. Measure the distance that each ball traveled before it hit the ground.

3. Weigh each ball.

4. Fill in the table showing how much each ball weighed and how far it traveled.

**Materials**
- balls of differing weights
- weighing scale
- measuring tape

| Ball | Weight | Distance traveled |
|------|--------|-------------------|
| #1   |        |                   |
| #2   |        |                   |
| #3   |        |                   |
| #4   |        |                   |

## Drawing Conclusions

Which ball traveled the greatest distance? Why?

_____
_____

Unit F · Motion and Energy    Use with TE textbook page F17

Name_____ Date_____

# Racing Balloon Rockets

**Quick Lab — FOR SCHOOL OR HOME — Lesson 2**

**Hypothesize** What forces work to make a balloon rocket go? Test your ideas. Write a **Hypothesis**:

_____
_____
_____

**Materials**
- soda straw
- tape
- balloon
- string

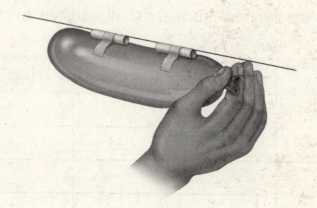

## Procedure

1. Look at the picture of how to construct a balloon rocket. Thread several pieces of soda straw onto the string. Then stretch the string tightly between two chairs.

2. **Observe** Blow up a balloon. Hold the neck closed with your fingers while your partner tapes two of the straw pieces to the balloon. Let go of the balloon, and watch what it does! Record your observations.

_____
_____

3. **Observe** How does the direction in which the balloon moves compare with the direction in which the air is forced out?

_____
_____

Name_____ Date_____

**QUICK LAB**
**FOR SCHOOL OR HOME**
**Lesson 2**

4. **Infer** Is there an unbalanced force on the balloon? In which direction does it push?

_____

_____

5. **Going Further** Think of your own questions that you might like to test. How would adding a second balloon affect your rocket's distance? Why?

My Question Is:

_____

_____

How I Can Test It:

_____

_____

_____

My Results Are:

_____

_____

Unit F · Motion and Energy  Use with textbook page F22

Name _____ Date _____

**Explore Activity**
**Lesson 3**

# Does Weight Affect How Fast an Object Falls?

**Hypothesize** Do objects of different weights fall at the same speed? Test your ideas.

Write a **Hypothesis**:

_____

_____

**Materials**
- table tennis ball
- golf ball
- pencil
- eraser
- goggles

**Procedure** **BE CAREFUL!** Wear goggles.

1. **Predict** Do heavy objects fall faster than lighter objects? Record your prediction and your reasons for making it.

   _____

   _____

   _____

   _____

2. **Observe** Stretch out your arms in front of you at shoulder height. Hold the two different balls—one in each hand—at the same height and drop them at exactly the same time. Listen for them to hit the floor. Which one hit the floor first? Record your results.

   _____

   _____

3. **Experiment** Repeat step 2 several more times to be sure your observations are accurate. Try dropping a pencil or an eraser at the same time as one of the balls. Record your observations.

   _____

   _____

Name_____ Date_____

## Drawing Conclusions

1. **Observe** Which ball hit the ground first?
   _____
   _____

2. **Observe** When you dropped different objects, which hit first, the heavier or the lighter?
   _____
   _____

3. **Hypothesize** Suggest an explanation for what you observed.
   _____
   _____
   _____

4. **FURTHER INQUIRY** **Experiment** Take two pieces of paper. Wad one into a tight ball. Leave the other alone. When you drop the two pieces of paper as you did the golf ball and table tennis ball, which will hit the ground first? Test your prediction. Explain your results.
   _____
   _____

### Inquiry

Think of your own question that you might like to test. Would the results change if you dropped a very light object, such as a paperclip, with one of the balls?

My Question Is:
_____
_____

How I Can Test It:
_____

My Results Are:
_____
_____

Name_____  Date_____

**Alternative Explore**
**Lesson 3**

# Pendulum

## Procedure

1. Make a pendulum by hanging a weight from the string.
2. Start swinging the pendulum back and forth. Then let gravity take over.
3. Now count the number of swings there are in one minute.
4. Repeat the experiment using different weights. Measure the mass of each weight. Keep all factors the same for each experiment.
5. Record your results in the table.

### Materials

- string
- weights
- stand for making pendulum

| Weight | Number of swings per minute |
|---|---|
| #1 | |
| #2 | |
| #3 | |
| #4 | |

## Drawing Conclusions

How does mass affect the results?

_____
_____
_____
_____
_____
_____
_____
_____

# Use Numbers

## What Do I Weigh on Other Worlds?

The Sun, planets, and moons in the solar system have different masses and radii. This causes the force of gravity at their surfaces to vary from world to world (for a gaseous planet, the "surface" is the top of its atmosphere). As the mass of any world increases, surface gravity tends to be stronger. However, as the radius increases, surface gravity tends to weaken. How would your weight change from one world to the next?

Table 1 lists gravity multipliers for solar system bodies. These values show the combined effect of the objects' different masses and radii on surface gravity compared with Earth. You can use the gravity multipliers to find your weight on other worlds. Just multiply your weight on Earth by the gravity multiplier for the new world. On Neptune, for example, your weight would be your weight on Earth multiplied by 1.1.

**Table 1**

| Object | Gravity (Earth = 1) |
|---|---|
| Sun | 28 |
| Moon | 0.16 |
| Mars | 0.38 |
| Jupiter | 2.6 |
| Saturn | 1.07 |
| Neptune | 1.1 |
| Venus | 0.91 |
| Mercury | 0.38 |
| Uranus | 0.91 |

**Table 2**

| World | Weight of a 250-Pound Astronaut | Your weight in pounds |
|---|---|---|
| Sun | 7,000 lb | |
| Moon | | |
| Mars | 95 lb | |
| Jupiter | | |
| Saturn | | |
| Neptune | | |
| Venus | | |
| Mercury | | |
| Uranus | 227.5 lb | |

## Procedure

1. **Analyze** Study Tables 1 and 2. Look carefully to see how numbers were used in the examples in Table 2.
2. **Use Numbers** Complete Table 2.

Name _____ Date _____

**Inquiry Skill Builder**
**Lesson 3**

## Drawing Conclusions

1. **Predict** A student who weighs 95 pounds on Earth has a mass of about 43 kg. What would the student's mass be on each world listed in Table 2?

   _____
   _____
   _____

2. **Infer** Saturn has much more mass than Earth, but your weight on Saturn is about the same as on Earth. How is this possible?

   _____
   _____
   _____
   _____

Name_____ Date_____

**Explore Activity**
Lesson 4

# What Makes Sound?

**Hypothesize** What causes sound? Remember, sounds can be different. How could you build an instrument to test your ideas?

Write a **Hypothesis:**

_____
_____
_____

**Materials**
- wood or plastic ruler
- long rubber band
- plastic or foam cup
- clear tape
- ballpoint pen
- scissors
- goggles

**Procedure** **BE CAREFUL!** Wear goggles.

1. As you do this activity, observe how sounds are made and changed. Poke a hole in the bottom of the cup. Cut the rubber band. Insert one end into the hole. Make a knot in the end to keep it in place.

2. Tape the cup and the stretched rubber band securely to the ruler as shown.

3. **Observe** Hold the cup next to your ear. Pluck the rubber band. Watch a partner do the same thing. Record what you hear and see.

_____
_____
_____

4. **Experiment** Put one finger on the rubber band, hold it against the ruler, and then pluck it again. What happens to the sound?

_____
_____

Unit F · Motion and Energy      Use with textbook page F49

Name_____ Date_____

## Drawing Conclusions

1. **Infer** What did you observe that made your instrument work? How can you explain what makes sound?

   _____
   _____

2. What happened to the sound when you changed the rubber band with your finger? Explain why, based on your observations.

   _____
   _____
   _____
   _____

3. **FURTHER INQUIRY** **Predict** What do you think will happen to the sound if you stretch the rubber band tighter? Untape the end of the rubber band and pull it a bit tighter. Retape the end to the ruler. Repeat steps 3 and 4. How do the results compare with your prediction? Give reasons for what happened.

   _____
   _____
   _____

**Inquiry**

Think of your own questions that you might like to test. What other factors might affect sound?

My Question Is:

_____
_____

How I Can Test It:

_____
_____

My Results Are:

_____

Name_____ Date_____

# Make a Drum

## Procedure

**Materials**
- balloon
- scissors
- rubber bands
- sturdy 8-oz plastic bowl
- paper scraps

1. Cut a piece of the balloon large enough to stretch over the top of the bowl. Hint: Make a cut in the balloon from the hole to the top.
2. While your partner holds the balloon across the top of the bowl, use a rubber band to hold the balloon in place.
3. Tap the balloon with the eraser end of a pencil. Describe what you hear.

   _____

4. Drop some paper scraps on the balloon. Repeat step 3. Describe what happens.

   _____

5. Tap the balloon with different amounts of force. Describe what happens.

   _____
   _____

## Drawing Conclusions

1. What happens to the paper scraps when you tap the balloon "drum?" What does this tell you about what happens to the balloon "drum" when you tap it?

   _____
   _____

2. What happens when you change the force with which you tap the balloon "drum?" What does this tell you about the amount of energy in the balloon "drum?"

   _____
   _____

3. Trace the flow of energy that begins when you tap the balloon.

   _____
   _____

Unit F · Motion and Energy     Use with TE textbook page F49

Name_____ Date_____

# Sound Carriers

**Hypothesize** Can sound travel through solids? Liquids?
Write a **Hypothesis**:
_____
_____

### Materials
- sealable pint-sized plastic food bag filled with water
- wind-up clock
- wooden table or desk

## Procedure

1. **Observe** Put a wind-up clock on a wooden table. Put your ear against the table. Listen to the ticking. Lift your head. How loud is it now?

_____
_____
_____

2. **Use Variables** Fill a sealable pint-size plastic bag with water. Seal the bag. Hold it against your ear. Hold the clock against the bag. How well can you hear the ticking? Move your ear away from the bag. How loud is the ticking now?

_____
_____
_____

Name_____ Date_____

## Drawing Conclusions

3. **Interpret Data** Rate wood, air, and water in order from best sound carrier to worst.
   _____

4. **Experiment** How would you test other materials, like sand?
   _____
   _____

5. **Going Further** Think of your own questions that you might like to test. Do some solids carry sound better than others?

   My Question Is:

   _____
   _____

   How I Can Test It:

   _____
   _____
   _____

   My Results Are:

   _____

Unit F · Motion and Energy          Use with textbook page F52

Name_____ Date_____

# How Can You Change a Sound?

**Hypothesize** Each musical instrument has a sound all its own. As you play an instrument, you make the sound change. What causes the sound to change? Test your hypothesis by building a homemade instrument from simple items like straws.

Write a **Hypothesis:**

_____
_____
_____

**Materials**
- 12 plastic drinking straws
- scissors
- metric ruler
- masking tape

## Procedure: Design Your Own

1. **Predict** Work in pairs to make a homemade instrument. Start with straws. Blow over one end of a straw. Will there be a difference if you seal the other end with tape? Record your prediction.

    _____
    _____

2. **Observe** Tape one end and blow over the open end. Describe what you hear. Does it work better with or without one end taped?

    _____
    _____

3. **Classify** Repeat with different lengths cut from a straw. Try at least four lengths. How are the sounds different? Arrange the straws in order to hear the difference.

    _____
    _____

4. **Experiment** Flatten one end of a straw. Cut the end to a point. Wet it. With your lips stretched across your teeth, blow into that end of the straw. Try to make different sounds with the straw. How might you modify the instrument the girl is using in the photograph on page F55 in your textbook?

Name_____ Date_____

## Drawing Conclusions

1. **Infer** Why do you think the sounds changed when you cut different lengths of straw? Hint: What is inside the straw—even if it looks empty?

   _____
   _____

2. **Communicate** Write a description of your instruments for a partner to build them exactly as you did. Include measurements taken with a ruler.

   _____
   _____
   _____

3. **FURTHER INQUIRY** **Experiment** Try other materials to make other instruments. Try such things as bottles with water, craft sticks, and so forth. Tell what causes the sound to change in each case.

   _____
   _____
   _____

### Inquiry

Think of your own questions that you might like to test. What other factors affect the sound an instrument makes?

My Question Is:

_____
_____

How I Can Test It:

_____
_____

My Results Are:

_____
_____

Unit F · Motion and Energy        Use with textbook page F55

Name_____ Date_____

**Alternative Explore — Lesson 5**

# Make Some Music

## Procedure

1. Look at each of the instruments to see how it is played.
2. Play each of the instruments. Describe the sound each one makes.

   _____
   _____

3. Choose one of the instruments and investigate ways to change the volume (make the sound louder or softer). Record your results.
4. Investigate ways to change the instrument's pitch (make the sound higher or lower). Record your results.
5. Repeat steps 3 and 4 for each of the other instruments. Use another sheet of paper if you need to add to the table.

**Materials**
- simple wind instruments, such as recorder, mouth organ, slide whistle, or kazoo

| Instrument | How to Change Volume | How to Change Pitch |
|---|---|---|
|  |  |  |
|  |  |  |
|  |  |  |
|  |  |  |

## Drawing Conclusions

1. What changes the volume of wind instruments?

   _____
   _____

2. What changes the pitch of wind instruments?

   _____
   _____
   _____

Name _____ Date _____

# Inquiry Skill Builder
### Lesson 5

# Communicate

**Making Tables and Graphs**

In this activity you will interpret data, classify sounds, and create your own table. Tables are helpful tools that organize information. The table shown gives the loudness of some common sounds in decibels (dB). Sounds below 30 dB can barely be heard. Quiet sounds are between 30 dB and 50 dB. Moderate sounds begin at 50 dB. At 70 dB, sounds are considered noisy. At 110 dB and above, sounds are unbearable.

| LOUDNESS OF SOME SOUNDS ||
|---|---|
| Sound | Loudness (in decibels) |
| Hearing limit | 0 |
| Rustling leaves | 10 |
| Whisper | 20 |
| Nighttime noises in house | 30 |
| Soft radio | 40 |
| Classroom/office | 50 |
| Normal conversation | 60 |
| Inside car on highway | 70 |
| Busy city street | 80 |
| Subway | 90 |
| Siren (30 meters away) | 100 |
| Thunder | 110 |
| **Pain threshold** | **120** |
| Loud indoor rock concert | 120 |
| Jet plane (30 meters away) | 140 |

## Procedure

1. **Classify** Determine which sounds are barely audible (can barely be heard), quiet, moderate, noisy, or unbearable.

   Barely audible (can barely be heard) _____

   _____

   Quiet _____

   _____

Unit F · Motion and Energy      Use with textbook page F59      229

Name _____ Date _____

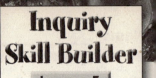

Moderate _____

_____

Noisy _____

_____

Unbearable _____

_____

2. **Communicate** Make your own table to show how you classified the sounds. Use another sheet of paper if needed.

3. **Communicate** Make a data table to record how many quiet, moderate, noisy, or unbearable sounds you hear in one hour. Make a graph to show your results. "Number" is the vertical axis. "Kind of Sound" is the horizontal axis. Put the data table and graph on a separate sheet of paper.

## Drawing Conclusions

1. **Interpret Data** How much louder is a soft radio than your house at night? A classroom than a house at night?

   _____

   _____

2. **Interpret Data** How much softer is normal conversation than thunder?

   _____

3. **Communicate** On another sheet of paper, make a chart listing loud sounds in the environment. What you can do to protect your ears from harm done by each loud noise?

   _____

   _____

Name_____ Date_____

**Explore Activity** — Lesson 6

# Do Sounds Bounce?

**Hypothesize** What happens when sound "hits" a surface? Does the kind of surface make a difference? Test your ideas.

Write a **Hypothesis:**

_____

_____

_____

### Materials
- 2 long cardboard tubes (can be taped, rolled-up newspapers)
- sound maker, such as a clicker or timer
- hard and soft test materials, such as a book, wood block, cloth, metal sheet, sponge, towel

## Procedure

1. Collect a variety of hard, smooth materials and soft, textured materials. Place one of the objects on a table. Set up your tubes in a V-shaped pattern on a table as shown in your textbook, page F65. The V should meet at the object you are testing. Record the name of the object in the first row of the table below.

2. **Observe** Place a sound maker (clicker or timer) at one end of the V. Listen for ticking at the other end of the V. Rank the loudness of the ticking on a scale of 1 (lowest) to 5 (highest). Record the number in the table.

3. **Experiment** Repeat steps 1 and 2 with the different materials you collected.

| Material/Object | Loudness Ranking |
|---|---|
| 1. | |
| 2. | |
| 3. | |
| 4. | |
| 5. | |

Unit F · Motion and Energy

Name_____ Date_____

## Drawing Conclusions

1. **Classify** What kinds of materials are the best reflectors—hard, smooth materials or soft, textured materials? What kinds of materials are the best absorbers?
   _____
   _____

2. **Make a Model** Draw a diagram of the path of sound from the sound maker to your ear. On your diagram mark the point in the path where the sound wave bounced.

3. **FURTHER INQUIRY** **Infer** Design an experiment to test the effectiveness of draperies or rugs in absorbing sound in a room.

### Inquiry

Think of your own questions that you might like to test. What other materials are good reflectors and absorbers?

My Question Is:
_____

How I Can Test It:
_____
_____

My Results Are:
_____
_____

232    Unit F · Motion and Energy    Use with textbook page F65

Name_____ Date_____

# Noisier or Quieter?

## Procedure

1. Work with a partner. Poke a hole in the bottom of each cup.

2. Thread a piece of string through each hole so that one end of the string is inside each cup. Knot the string and tape each end to the bottom of the cup.

3. You and your partner hold the cups so that the string is pulled tight. While you hold one cup to your ear, have your partner speak into the other cup, using a normal tone of voice. Describe what you hear.

_____
_____
_____

**Materials**
- 2 paper or plastic cups
- string
- tape
- tissues

4. Switch roles with your partner, so that your partner has a chance to hear, too.

5. Put some tissues in the your cup. Then repeat step 3. Describe what you hear.

_____
_____
_____

6. Switch roles with your partner, so that your partner has a chance to hear, too.

## Drawing Conclusions

1. When was the sound you heard quieter, with or without the tissues?

_____
_____

2. Explain why there was a difference.

_____
_____

Unit F · Motion and Energy    Use with TE textbook page F65

Name_____ Date_____

# Clap! Clap!

**QUICK LAB**
FOR SCHOOL OR HOME
Lesson 6

**Hypothesize** Can you cause a clear time lag between a sound and its echo? Write a **Hypothesis**:

_____

_____

### Materials
- meterstick

## Procedure

1. **Observe** Stand about 8 m away from a large wall, such as the side of your school building. Make sure there is plenty of open space between you and the wall. Clap your hands, and listen for an echo. Notice how much time there is between your clap and the echo.

_____

_____

_____

2. **Observe** Move closer to the wall, and clap again. Listen for an echo. Try this several times.

_____

_____

_____

Name_____ Date_____

## Drawing Conclusions

3. **Observe** As you got closer to the wall, how did the time between the clap and the echo change? Did you always hear an echo? Explain.

_____
_____
_____

4. **Experiment** Repeat at different distances. What happens?

_____
_____
_____

5. **Going Further** Think of your own questions that you might like to test. Will you get the same results if you try other sounds?

   My Question Is:

   _____

   How I Can Test It:

   _____
   _____
   _____

   My Results Are:

   _____
   _____

Name_____ Date_____

Lesson 7

# Can You See Without Light?

**Hypothesize** Is it possible to see objects if there is no light? Test your ideas.

Write a **Hypothesis:**

_____

_____

## Procedure

**BE CAREFUL!** Handle scissors carefully. Do not put any sharp objects in the box.

1. How could you design a test to determine how well you can see without light? Cut a hole about the size of a dime in the box as shown. Put an object inside the box. Close the lid.

2. **Observe** Look in the box through the hole. What do you see? Write a description of it.

_____

3. Now cut a small hole in the top of the box.

4. **Experiment** Shine the flashlight through the top hole while you look into the box again. Can you see the object this time?

_____

### Materials

- small cardboard box with lid
- small object to put inside box, such as an eraser, crayon, or coin
- scissors
- flashlight

## Drawing Conclusions

1. **Communicate** Could you see the object inside the box in step 2? In step 4? Explain any difference in your answers.

_____

_____

_____

2. **Infer** Is it possible to see an object in the dark? Explain.

_____

_____

_____

Name_____ Date_____

**Explore Activity** — Lesson 7

3. **Predict** Do any characteristics of the object in the box affect the results? Try different kinds of objects. Predict any differences in your results. Test your ideas.

_____
_____
_____
_____

4. **FURTHER INQUIRY** **Predict** How much extra lighting would you need on a dark, cloudy day in order to safely walk around your classroom or your room at home? Would a night-light work? How would you test your ideas safely?

_____
_____
_____
_____

**Inquiry**

Think of your own questions that you might like to test. What other factors about objects in the box might make them visible?

My Question Is:

_____

How I Can Test It:

_____

My Results Are:

_____

Name_____ Date_____

**Alternative Explore**
**Lesson 7**

# Can You See the Flashlight?

**Procedure** **BE CAREFUL!** Handle scissors carefully.

1. Cut a small hole in the end of a cardboard box. Place the flashlight inside the box. Do not turn on the flashlight.
2. Discuss with your group whether you can see the flashlight if you look through the hole.
3. Try it. Can you see the flashlight?
   _____
4. What could you do to be able to see the flashlight inside the box? Make a list of your group's suggestions.
   _____
   _____
   _____
5. Test your ideas. Record your observations.
   _____
   _____
   _____

**Materials**
- scissors
- flashlight
- small cardboard box with lid, such as a shoebox

## Drawing Conclusions

1. What ideas did your group test? Did they make it possible for you to see the flashlight?
   _____
   _____
   _____

2. What do your results tell you is necessary for you to be able to see something?
   _____
   _____

Name_____ Date_____

# Follow the Bouncing Light

**QUICK LAB** — FOR SCHOOL OR HOME — Lesson 7

**Hypothesize** How does light travel when it bounces off a mirror? Write a **Hypothesis**:

_____

_____

### Materials
- mirror
- string

## Procedure

1. Hold a small pocket mirror as shown. Adjust it so your partner can see your face in the middle of the mirror.

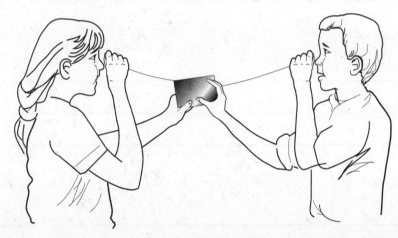

2. You and your partner should hold a long string taut to the mirror at the point where you see each other's nose. Compare the two angles formed between the string and the mirror.

3. **Observe** Move a little farther apart. How does the mirror have to be moved for your partner to see your face?

_____

_____

_____

Name_____ Date_____

**Quick Lab** — FOR SCHOOL OR HOME — Lesson 7

## Drawing Conclusions

4. **Interpret Data** What did you observe about the angles the string made with the mirror?

   _____
   _____
   _____

5. **Going Further** Think of your own questions that you might like to test. Will light bounce off objects other than mirrors?

   My Question Is:

   _____

   How I Can Test It:

   _____
   _____
   _____

   My Results Are:

   _____
   _____
   _____

Name_____ Date_____

# What Can Light Pass Through?

**Hypothesize** How do objects cast shadows? Do all objects cast shadows the same way? Are all shadows alike? How would you test your ideas?

Write a **Hypothesis:**

_____
_____
_____

**Materials**
- plastic sandwich bag
- paper
- waxed paper
- aluminum foil
- other assorted materials to test
- flashlight
- clear-plastic cup
- water (other liquids, optional)
- food dye

## Procedure

1. **Classify** Sort the test materials into those that you think light can pass through and those that light cannot pass through.

2. **Experiment** Use the flashlight to test if light can pass through each of the solid materials. Record your observations. Test if light will pass through water. What about water colored with food dye?

_____
_____
_____
_____

3. **Infer** How can you test if light passes through gases? Explain. What materials would you need?

_____
_____

## Drawing Conclusions

1. **Interpret Data** Can light pass through all the materials equally well?

_____

Name _____ Date _____

**Explore Activity**
**Lesson 8**

2. **Interpret Data** Can light pass through solids, liquids, and gases?

   _____
   _____

3. **Predict** What else might you add to water to see if light gets through—sand, ink, instant coffee? Predict if each lets light through. How would you test your ideas?

   _____
   _____
   _____

4. **FURTHER INQUIRY** **Experiment** Design a room from window coverings to lighting, where shadows of objects are always soft and fuzzy, never sharp. What sorts of materials would you use?

   _____
   _____
   _____

**Inquiry**

Think of your own questions that you might like to test. What other factors affect shadows?

My Question Is:

_____

How I Can Test It:

_____
_____

My Results Are:

_____

Name_____ Date_____

**Alternative Explore**
**Lesson 8**

# Rank Shadows

## Procedure

1. Sort through the materials your teacher gives you to work with. Predict which will make the sharpest shadows. Record your predictions.

   _____
   _____

2. With the room somewhat darkened, shine a flashlight on each item to be tested. Try to make a shadow on a wall. Notice how sharp each shadow is. Record your observations.

   _____
   _____

### Materials
- plastic sandwich bag
- aluminum foil
- waxed paper
- flashlight
- clear-plastic cup
- water
- food dye
- other materials to test
- paper

## Drawing Conclusions

1. Rank the materials in terms of the shadows they produce, from the sharpest shadow to the least sharp.

   _____
   _____
   _____
   _____

2. What pattern can you see in the kinds of materials you tested and the shadows they made?

   _____
   _____

Unit F · Motion and Energy     Use with TE textbook page F95

Name_____ Date_____

# Seeing Through a Lens

**QUICK LAB**
FOR SCHOOL OR HOME
Lesson 8

**Hypothesize** What happens when you view the room through a lens? How does it change the way things look? Write a **Hypothesis:**

_____

_____

### Materials

- convex lens (magnifying glass)
- index card or piece of paper

## Procedure

1. **Observe** Hold a convex lens about a foot from your eye. View the image of the room around you. Record what you see. Repeat with the lens quite close to the page of a book.

   _____
   _____
   _____
   _____

2. **Experiment** Aim the lens at a light bulb or window. Move an index card back and forth on the other side of the lens until you see an image of the light source cast sharply on the card. Record what you see.

   _____
   _____

Name_____ Date_____

**Quick Lab**
FOR SCHOOL OR HOME
Lesson 8

## Drawing Conclusions

3. **Observe** When the image was upright, was it enlarged or reduced?
   _____

4. **Observe** When you cast an image on the card, was it upright or inverted?
   _____

5. **Classify** Summarize your observations in a table on a separate sheet of paper.
   _____

6. **Going Further** Think of your own questions that you might like to test. Does the distance of the object from the lens affect the image?

   My Question Is:
   _____
   _____

   How I Can Test It:
   _____
   _____

   My Results Are:
   _____

Name_____ Date_____

**Explore Activity**
Lesson 9

# What Is Color?

**Hypothesize** What color will a blue object appear to be if you look at it under a blue light? Under a red light? How could you test your ideas even if you did not have a red or blue light bulb?

Write a **Hypothesis:**

_____
_____
_____

**Materials**

- red, yellow, blue, and green cellophane sheets

- white paper

- crayons

- red, yellow, blue, green, and black squares of construction paper

- flashlight

## Procedure

1. **Observe** Instead of using colored light bulbs, shine a flashlight at a sheet of white paper through each of the cellophane sheets. Record what you see.

   _____

2. **Predict** Is there a difference if you observe the paper by looking through colored cellophane instead? What color will each of the colored squares appear to be through each of the cellophane sheets? Check your predictions.

   _____
   _____
   _____
   _____

3. **Make a Model** Use the crayons to make additional colored squares to view through the cellophane sheets.

4. **Communicate** Make a table on a separate sheet of paper that shows what color each square appears to be through each of the cellophane sheets.

246    Unit F · Motion and Energy    Use with textbook page F107

Name_____ Date_____

## Drawing Conclusions

1. **Communicate** What color does the red square appear to be when viewed through the red cellophane sheet? Why? What color does the blue square appear to be when viewed through the red cellophane sheet? Why?

_____
_____
_____
_____
_____

2. **FURTHER INQUIRY** **Predict** What do you think would happen if you looked at the red square through both the red and blue cellophane sheets at the same time? Try it to test your prediction.

_____
_____
_____

## Inquiry

Think of your own questions that you might like to test. How does mixing colors of light differ from other ways to mix colors?

My Question Is:

_____

How I Can Test It:

_____
_____

My Results Are:

_____
_____

Name_____ Date_____

# What Is Color?

**Procedure** BE CAREFUL! Handle scissors carefully.

1. Cut a rectangle from the lid of the box. Tape red cellophane over the hole.
2. Cut a small viewing hole in the side of the box.
3. Place a small white object in the box. Replace the lid.
4. Shine a flashlight through the cellophane. Look at the object through the viewing hole. What color does the object appear to be?

   _____

5. Remove the white object and place a red object and a green object in the box. Replace the lid.
6. Shine a flashlight through the cellophane. Look at the objects through the viewing hole. What color does each object appear to be?

   _____
   _____

**Materials**
- shoe box with lid
- scissors
- red cellophane
- tape
- small white object
- small red object
- small green object
- flashlight

## Drawing Conclusions

1. What color did the white object look inside the box? _____
2. What color did the red object look inside the box? _____
3. What color did the green object look inside the box?

   _____

4. Explain your results.

   _____
   _____
   _____
   _____

Name_____ Date_____

**Inquiry Skill Builder**
Lesson 9

# Predict

## Mixing Colors

You will use pigments–colored substances–in this activity to see the way pigments blend to make the other colors.

In this activity you will make a prediction before you do the activity. That is, you will make a reasonable guess about what you expect the results to be. Predict what colors will result when you mix certain colors of food dyes together.

**Materials**
- red, yellow, blue and green food dyes
- water
- plastic cups
- goggles

**Procedure** BE CAREFUL! Wear goggles.

1. Place four cups on a piece of paper. Add enough water to each cup to cover the bottom.

2. **Predict** What color will be made by mixing one drop of red food dye and one drop of yellow food dye in the water? Mix well. Record the result.

| Prediction | |
|---|---|
| Results | |

3. **Experiment** Do step 2 with red and blue dyes. Be sure to make a prediction before you mix the colors.

| Prediction | |
|---|---|
| Results | |

Name_____ Date_____

**Inquiry Skill Builder**
**Lesson 9**

4. **Experiment** Do step 2 again with yellow and blue, and then with all four colors. Again, be sure to make your predictions before you mix the colors.

| Prediction | |
|---|---|
| Results | |

| Prediction | |
|---|---|
| Results | |

**Drawing Conclusions**

1. **Communicate** What color resulted when you mixed red and yellow?

   _____
   _____

2. **Communicate** What color resulted when you mixed red and blue? Blue and yellow? When you mixed all four colors?

   _____
   _____

3. **Infer** What would happen if you used different amounts of each dye? Experiment to find out. Make predictions about the final color before you mix the dyes.

| Color/Amount | Prediction | Results |
|---|---|---|
| | | |
| | | |
| | | |
| | | |

250  Unit F · Motion and Energy  Use with textbook page F111

Name_____ Date_____

**Explore Activity**
Lesson 10

# How Do Waves Move?

**Hypothesize** How can you make waves move faster or slower? Test your ideas.

Write a **Hypothesis:**

_____

_____

_____

**Materials**
- spring toy
- meterstick
- stopwatch or digital watch

## Procedure

1. One way to experiment with waves is to use a spring toy as a model. Work in groups of three. Two students should stretch the spring toy out 2 meters. One student should jiggle the spring toy slowly up and down to form waves that move along its length.

2. **Observe** The third student should time how long the wave takes to travel from end to end. Repeat several times. Record the results.

3. **Experiment** See what factors affect the size and speed of the wave produced. Compare results when the spring toy is loosely stretched and tightly stretched.

| Description of Model | Time |
|---|---|
|  |  |
|  |  |
|  |  |
|  |  |

Name_____ Date_____

**Explore Activity**
Lesson 10

## Drawing Conclusions

1. **Observe** In what direction does the wave move? In what direction do the spirals move?

   _____

   _____

2. **Interpret Data** How does holding it tighter or looser change how the wave moves?

   _____

   _____

3. **FURTHER INQUIRY** **Experiment** Try moving one end of the spring toy with a faster speed of the up and down movement. Again, vary the length of the spring toy. What happens?

| Paper-Clip Combination | Results |
|---|---|
|  |  |
|  |  |
|  |  |

**Inquiry**

Think of your own questions that you might like to test. Does the strength of the force affect the motion of a wave?

My Question Is:

_____

How I Can Test It:

_____

_____

My Results Are:

_____

252    Unit F · Motion and Energy    Use with textbook page F115

Name _____ Date _____

**Alternative Explore**
Lesson 10

# Change of Speed

## Procedure

1. Work in a group. Two students will hold the ends of the rope. A third student will be the counter. A fourth student will be the recorder.

2. The students holding the rope should hold it taut. One student starts a single vertical vibration at one end while the counter begins to count seconds or keep time on a watch with a second hand. The counter should stop as soon as the vibration reaches the other end of the rope.

3. Record the time it took for the wave to travel the length of the rope.

   _____

4. Group members should change jobs and repeat the experiment.

5. Repeat the experiment with the thinner rope. Record the time it took for the wave to travel the length of the rope.

   _____

6. Try varying the height of the vibration on each rope. Does this change the time it takes for the wave to move along the rope?

   _____

### Materials

- jump rope
- thinner rope

## Drawing Conclusions

1. On which rope did the wave move more quickly?

   _____

2. Did the height of the vibration that moved down the rope affect the speed?

   _____

3. What determines the speed at which a wave travels?

   _____

   _____

Unit F · Motion and Energy   Use with TE textbook page F115

Name_____ Date_____

# Water Waves

**Quick Lab**
**FOR SCHOOL OR HOME**
Lesson 10

**Hypothesize** How do water waves affect the motion of floating objects? Test your ideas. Write your **Hypothesis:**

_____

_____

### Materials

- aluminum foil
- shallow pan at least 20 x 28 cm
- water
- pencil

## Procedure

1. Fill a shallow pan or tray (20 by 28 cm) half full of water. Fold small squares of foil (1 cm by 1 cm) into tiny "boats." Place several of these boats on the surface of the water.

2. At one end of the tray, make waves on the water's surface. Do this by moving your pencil horizontally up and down in the water.

3. **Predict** What do you think will happen to the boats after 30 seconds? After one minute? Record your predictions.

_____

_____

254 | Unit F · Motion and Energy | Use with textbook page F117

Name_____ Date_____

## Drawing Conclusions

4. **Observe** What happened to the boats? How did they move? How far did they move? Were your predictions correct?

   _____
   _____
   _____

5. **Experiment** What happens if you change how fast you make the waves? What happens if you change the number of boats you use?

   _____
   _____

6. **Going Further** Think of your own questions you might like to test. How do the waves produced by other objects move?

   My Question Is:

   _____

   How I Can Test It:

   _____
   _____
   _____

   My Results Are:

   _____

Unit F · Motion and Energy      Use with textbook page F117